D1228940

Microwave Synthesis

Chemistry at the Speed of Light

by Brittany L. Hayes, Ph.D.

Printed in U.S.A.

ISBN 0-9722229-0-1

Library of Congress Control Number: 2002109572

CEM Publishing
PO Box 200
Matthews, NC 28105

Contents

Foreword

By Alan Katritzky, Ph.D.

Virtually all types of thermally driven chemical reactions can be accelerated by microwaves: additions, cycloadditions, substitutions, eliminations, fragmentations...the list goes on. This book, written by a practicing organic chemist for practicing organic chemists, shows how this powerful technique can be applied in laboratory situations. It offers advice on the selection of conditions, solvents, temperatures. It explains the manifold possibilities of modern instrumentation, and provides a powerful tool for the practical application of the technique to enhance the range of synthesis and the productivity of synthetic chemists.

Although microwave-enhanced organic chemistry has been around since the mid-1980s, until recently, it frequently utilized domestic microwave ovens and consequently lacked control and reproducibility, thus for the most part, it was unsafe. Now, in large part due to the development of efficient technology, it is rapidly gaining acceptance and popularity. The number of publications on microwave-assisted organic chemistry is increasing exponentially, as it is being realized that microwave systems provide the opportunity to complete reactions in minutes, and have manifold applications in academic and industrial environments alike.

With the advent of new single-mode technology at a reasonable cost, together with the simplicity of its use and installation in a normal synthetic laboratory, chemists now have access to greater enhancements, feedback control, and repeatability. Available commercial systems contain temperature and pressure sensors, built-in magnetic stirring, cooling mechanisms, power control, software operation, and even automation. Affordable instruments are safe, reliable, and effective. Microwave instrumentation is as easy to install and operate as a hotplate...place it in a hood, plug it in, and you're on your way.

This book provides an educational tool which can benefit chemists at all levels from graduate students to senior level section leaders with much experience in synthetic chemistry. The well-known ability of microwave energy to increase the internal temperatures of organic substances more rapidly than conventional thermal heat can greatly profit synthetic chemistry. An understanding of the basics of microwave energy can be used in conjunction with general chemical knowledge to enhance synthesis. This book is the first to describe microwave chemistry from the point of view of an organic chemist and it will be much appreciated.

Alan Katritzky, Ph.D.
Kenan Professor of Chemistry and Director of the Center for Heterocyclic Compounds at the University of Florida

Chapter 1
Introduction to Microwave Chemistry

By Michael J. Collins, Ph.D.

Microwave synthesis represents a major breakthrough in synthetic chemistry methodology, a dramatic change in the way chemical synthesis is performed and in the way it is perceived in the scientific community. Conventional heating, long known to be inefficient and time-consuming, has been recognized to be creatively limiting as well. Microwave synthesis gives organic chemists more time to expand their scientific creativity, test new theories and develop new processes. Instead of spending hours or even days synthesizing a single compound, chemists can now perform that same reaction in minutes. In concert with a rapidly expanding applications base, microwave synthesis can be effectively applied to any reaction scheme, creating faster reactions, improving yields, and producing cleaner chemistries.

Microwave synthesis gives organic chemists more time to expand their scientific creativity, test new theories and develop new processes.

In addition, microwave synthesis creates completely new possibilities in performing chemical transformations.

Because microwaves can transfer energy directly to the reactive species, so-called “molecular heating”, they can promote transformations that are currently not possible using conventional heat. This is creating a new realm in synthetic organic chemistry.

Microwaves also provide chemists with the option to perform “cool reactions”. Energy is applied directly to the reactants. However, the bulk heating is minimized by use of simultaneous cooling. This allows for enhanced reactions of larger, more heat sensitive molecules (e.g. proteins), as the temperatures are low enough to eliminate thermal degradation. This will provide some exciting new opportunities and an important new tool for proteomics and genomics research.

Recent microwave hardware advancements now provide a range of affordable, flexible tools for the synthetic chemist. This new technology, coupled with the rapidly expanding knowledge and applications base, will cause a major shift towards microwave synthesis in the next few years. As Victor Hugo, the famous French novelist and poet wrote, “An invasion of armies can be resisted, but not an idea whose time has come.” Microwave synthesis is an idea whose time has come and whose impact will be truly monumental on the world of chemistry.

History

The development of microwave technology was stimulated by World War II, when the magnetron was designed to generate fixed frequency microwaves for RADAR devices.[1,2] Percy LeBaron Spencer of the Raytheon Company accidentally discovered that microwave energy could cook food when a candy bar in his pocket melted while he was experimenting with radar waves. Further investigation showed that microwaves could increase the internal temperature of foods much quicker than a conventional oven. This ultimately led to the introduction of the first commercial microwave oven for home use in 1954.

Investigation into the industrial applications for microwave energy also began in the 1950s and has continued to the present. Microwave energy has found many uses including irradiating coal to remove sulfur and other pollutants, rubber vulcanization, product drying, moisture and fat analysis of food products, and solvent extraction applications. Wet ashing or digestion procedures for biological and geological samples have also become very important analytical tools. As improvements and simplifications were made in magnetron design, the prices of domestic ovens fell significantly. Consequently, research done in the latter half of the 20th century was performed in modified domestic microwave ovens. The effects of microwave irradiation in organic synthesis were not explored until the mid 1980s. The first two papers on microwave-enhanced organic chemistry were published in 1986 and many organic chemists have since discovered the benefits of using microwave energy to drive synthetic reactions.[3,4] Until recently, most of this research has been executed in multi-mode domestic microwave ovens, which have proven to be problematic. These ovens are not designed for the rigors of laboratory usage: acids and solvents corrode the interiors quickly; there are no safety controls, temperature or pressure monitoring; and the cavities are not designed to withstand the resulting explosive force from a vessel failure in runaway reactions.

In the 1980s, companies began to address these issues by manufacturing industrial microwave ovens specifically designed for use in laboratories. These multi-mode systems featured corrosion-resistant stainless steel cavities with reinforced doors, temperature and pressure monitoring, and automatic safety controls.

They have worked well for doing large-scale laboratory applications, but they have some fundamental limitations in performing small-scale synthetic chemistry. Recently, single-mode technology, which provides more uniform and concentrated microwave power, has become available. These newer systems represent a breakthrough in providing new capabilities for doing microwave synthesis and are a key factor in the rapid expansion of this field of science.

Microwave Theory

Microwaves are a powerful, reliable energy source that may be adapted to many applications. Understanding the basic theory behind microwaves will provide the organic chemist with the right tools and knowledge to be able to effectively apply microwave energy to any synthetic route.

Figure 1

The electromagnetic spectrum

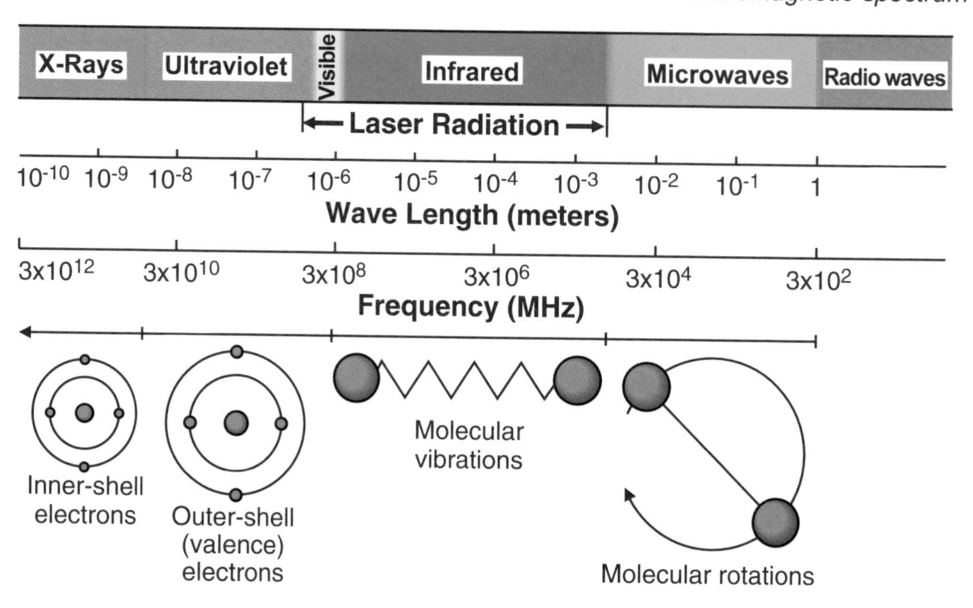

I. What are microwaves?

A **microwave** is a form of electromagnetic energy that falls at the lower frequency end of the electromagnetic spectrum, and is defined in the 300 to about 300,000 megahertz (MHz) frequency range (Figure 1). Within this region of electromagnetic energy, only molecular rotation is affected, not molecular structure.[1] Out of four available frequencies for industrial, scientific, or medical applications, 2450 MHz is preferred because it has the right penetration depth to interact with laboratory scale samples, and there are power sources available to generate microwaves at this frequency.

Microwave energy consists of an electric field and a magnetic field, though only the electric field transfers energy to heat a substance (Figure 2).[1] Magnetic field interactions do not normally occur in chemical synthesis.

Figure 2

A microwave

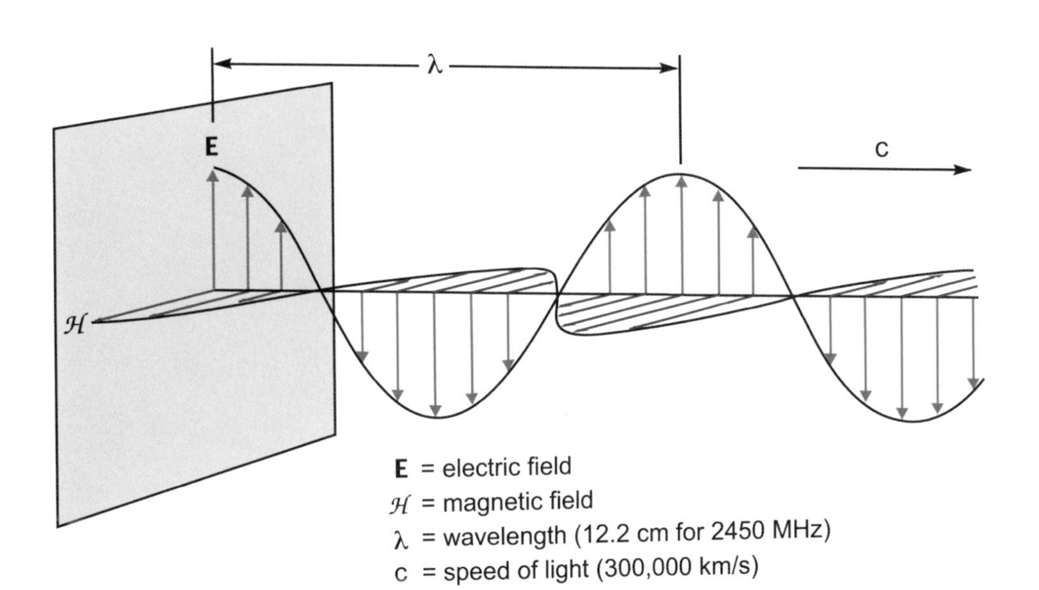

Microwaves move at the speed of light (300,000 km/sec). The energy in microwave photons (0.037 kcal/mole) is very low relative to the typical energy required to cleave molecular bonds (80-120 kcal/mole); thus, microwaves will not affect the structure of an organic molecule. In the excitation of molecules, the effect of microwave absorption is purely kinetic.

II. How does a microwave heat a substance?

Traditionally, chemical synthesis has been achieved through conductive heating with an external heat source. Heat is driven into the substance, passing first through the walls of the vessel in order to reach the solvent and reactants (Figure 3). This is a slow and inefficient method for transferring energy into the system because it depends

Figure 3

Schematic of sample heating by conduction

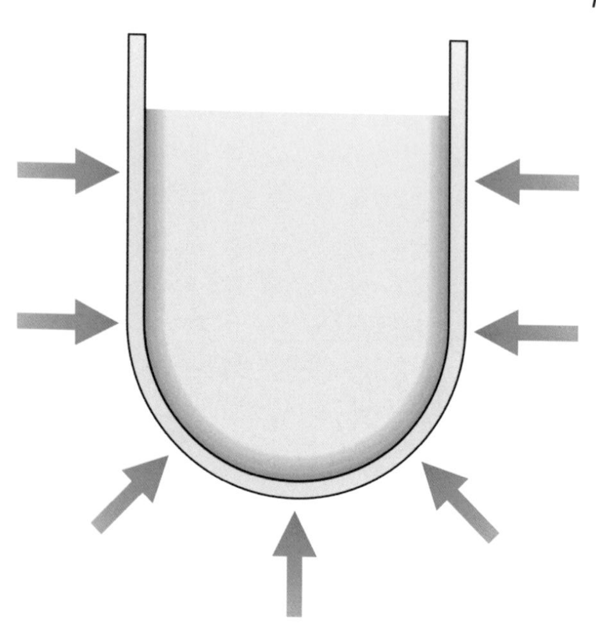

Temperature on the outside surface is greater than the internal temperature.

on the thermal conductivity of the various materials that must be penetrated. It results in the temperature of the vessel being higher than that of the reaction mixture inside until sufficient time has elapsed to allow the container and contents to attain thermal equilibrium. This process can take hours. Conductive heating also hinders the chemist's control over the reaction. The heat source must physically be removed and cooling administered to reduce the internal bulk temperature.

Microwave heating, on the other hand, is a very different process. As shown in Figure 4, the microwaves couple directly with the molecules that are present in the reaction mixture, leading to a rapid rise in temperature. Because the process is not dependent upon the thermal conductivity of the vessel materials, the result

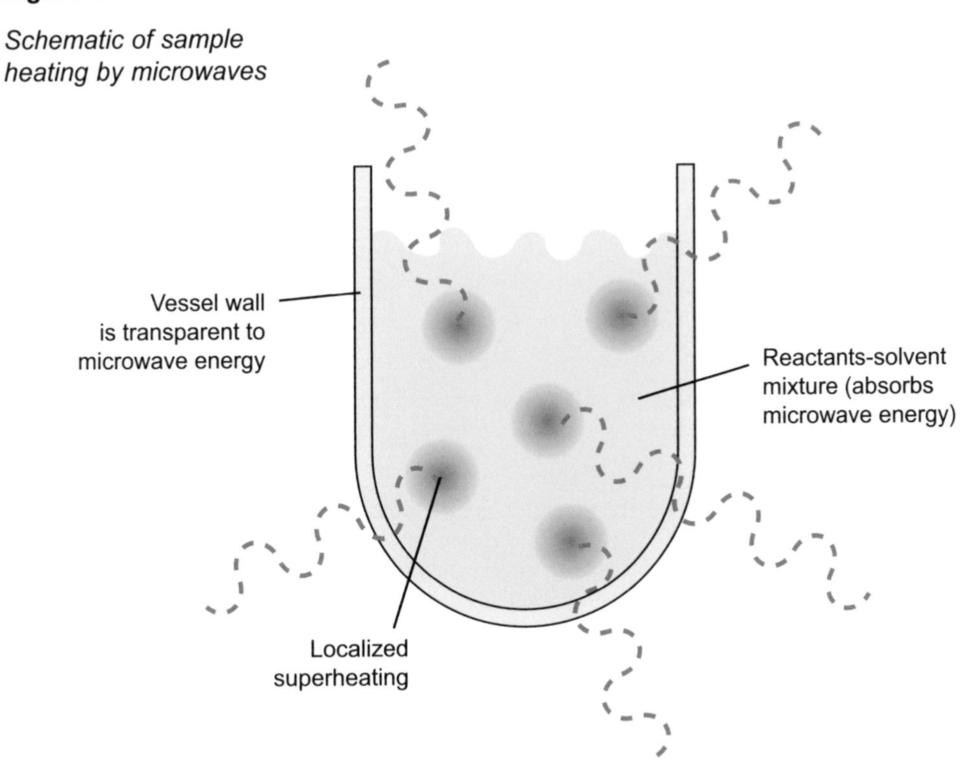

Figure 4

Schematic of sample heating by microwaves

is an instantaneous localized superheating of anything that will react to either **dipole rotation** or **ionic conduction**, the two fundamental mechanisms for transferring energy from microwaves to the substance being heated. Microwave heating also offers facile reaction control. It can be described as "**instant on-instant off**". When the microwave energy is turned off, latent heat is all that remains.

Dipole rotation is an interaction in which polar molecules try to align themselves with the rapidly changing electric field of the microwave. The rotational motion of the molecule as it tries to orient itself with the field results in a transfer of energy. The coupling ability of this mechanism is related to the polarity of the molecules and their ability to align with the electric field. There are a number of factors that will ultimately determine the dipole rotation coupling efficiency; however, any polar species (solvent and/or substrate) that are present will encounter this mechanism of energy transfer.

The second way to transfer energy is ionic conduction, which results if there are free ions or ionic species present in the substance being heated. The electric field generates ionic motion as the molecules try to orient themselves to the rapidly changing field. This causes the instantaneous superheating previously described. The temperature of the substance also affects ionic conduction: as the temperature increases, the transfer of energy becomes more efficient.

III. How do microwaves increase reaction rates?

In a typical reaction coordinate (Figure 5), the process begins with reactants (A and B), which have a certain energy level (E_R). In order to complete the transformation, these reactants must collide in the correct geometrical orientation to become activated to a higher-level transition state (E_{TS}). The difference between these energy levels is the activation energy (E_a) required to reach this higher state ($E_{TS} - E_R = E_a$). The activation energy is the energy that the system must absorb from its environment in order to react. Once enough energy is absorbed, the reactants quickly react and return to a lower energy state (E_P) — the products of the reaction (A-B). Microwave irradiation does not affect the activation energy, but provides the momentum to overcome this barrier and complete the reaction more quickly than conventional heating methods.

Figure 5

Reaction coordinate

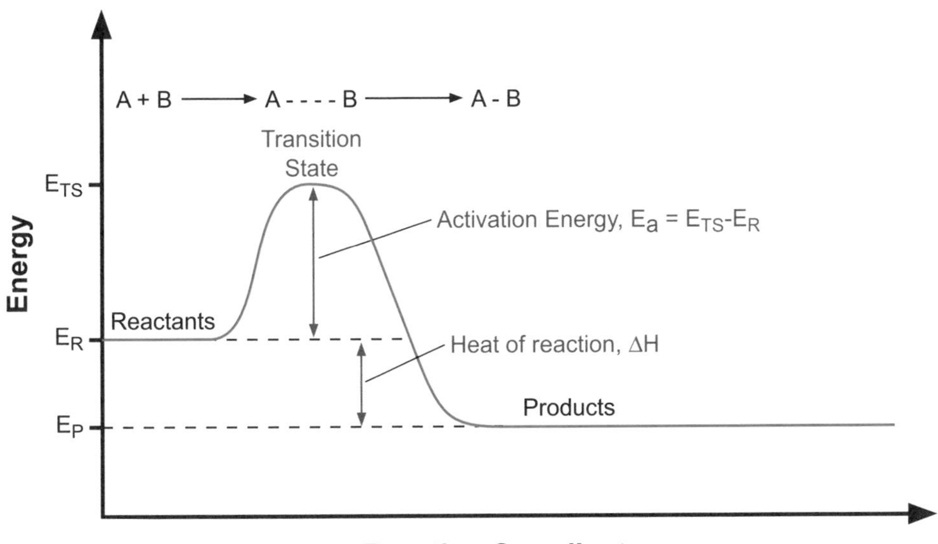

In general drug discovery and development work, for example, an activation energy value may be 50 kcal/mole. A representative process would involve 30 mg of each reactant, leading to approximately 30 mg of product(s) with an average molecular weight of 300 g/mole (300-350 is characteristic of present drug compounds). According to the calculations in Figure 6, five calories of energy are required for the full transformation to occur. Commercial single-mode microwave hardware available today typically delivers 300 W of power. Translated into calories, this indicates that 72 cal/sec of energy are available from the 300 W of microwave power, assuming a 100% efficiency in microwave heating. Clearly, the amount of microwave energy being introduced to the system is very large relative to the energy needed to achieve the required activation energy. It is this phenomenon that contributes to the increased reaction speeds and higher yields that occur in microwave chemistry.

Figure 6

Microwave energy vs. required activation energy

$$0.03\text{ g} \times \frac{\text{mol}}{300\text{ g}} \times \frac{50000\text{ cal}}{\text{mol}} = \mathbf{5\ cal}$$

product — molecular weight — activation energy

$$300\text{ W} = \frac{300\text{ J}}{\text{sec}} \times \frac{0.239\text{ cal}}{\text{J}} = \mathbf{72\ cal/sec}$$

microwave energy — conversion factor

One of the most important aspects of microwave energy is the rate at which it heats (Figure 7). Microwaves will transfer energy in 10^{-9} seconds with each cycle of electromagnetic energy. The kinetic molecular relaxation from this energy is approximately 10^{-5} seconds. This means that the energy transfers faster than the molecules can relax, which results in the non-equilibrium condition and high instantaneous temperatures that affect the kinetics of the system. This, in turn, enhances the reaction rate, as well as the product yields. In addition, the lifetime of activated complexes are approximately 10^{-13} seconds, and thus, are much shorter lived than the rate at which energy is transferred with microwaves. Activated complexes do not normally exist long enough to have an opportunity to absorb microwave energy. However, there are a number of resonance-stabilized intermediates that are much longer lived. Many of these have lifetimes longer than 10^{-9} seconds, so the opportunity exists in certain chemical reactions, for intermediates generated in this approximate time frame, to couple directly with the microwave and be further enhanced. Most intermediates are highly polar species and many of them are even ionic in character, making them excellent candidates for microwave energy transfer.

Figure 7

Speed of microwave heating

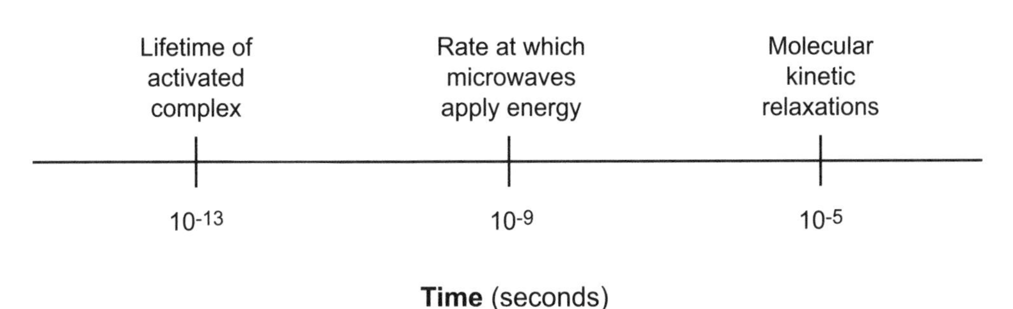

Based on the **Arrhenius reaction rate equation ($k = Ae^{-E_a/RT}$)**, the reaction rate constant is dependent on two factors: the frequency of collisions between molecules that have the correct geometry for a reaction to occur (A) and the fraction of those molecules that have the minimum energy required to overcome the activation energy barrier ($e^{-E_a/RT}$). There has been some speculation that microwaves affect the orientation of the molecular collisions and the activation energy, but there is no evidence that supports either of these ideas.[5,6] Microwaves do not influence the orientation of those collisions, nor the activation energy. Activation energies remain constant for each particular reaction. However, microwave energy will affect the temperature parameter in this equation. An increase in temperature causes molecules to move about more rapidly, which leads to a greater number of more energetic collisions. This occurs much faster with microwave energy, due to the high instantaneous heating of the substance(s) above the normal bulk temperature, and is the primary factor for the observed rate enhancements.

Arrhenius reaction rate equation

$$(k = Ae^{-E_a/RT})$$

It should also be obvious that the level of instantaneous heating will be dependent on the amount of microwave energy that is used to irradiate the reactants. The higher the level of microwave energy, the higher the instantaneous temperature will be relative to the bulk temperature. One method for increasing the microwave energy that is delivered is to use simultaneous cooling during the microwave irradiation. This allows a higher level of microwave power to be directly administered, but will prevent overheating by continuously removing latent heat. This technique has proven to be very effective in further enhancing of reaction rates and will be discussed in greater detail throughout the book.

Based on experimental data from numerous works that have been performed over the last ten years, chemists have found that microwave-enhanced chemical reaction rates can be faster than those of conventional heating methods by as much as 1,000-fold.[3,4,7-15] Assuming the standard first order rate law (rate = $k[A]$), the Arrhenius rate equation ($k = Ae^{-E_a/RT}$) was used to calculate the instantaneous temperatures required to get the reaction enhancements shown in Figure 8. The assumption was a desired reaction bulk temperature of 150 °C and an activation energy of 50 kcal/mole for the transformation. For a 10-fold rate increase, it was determined that a temperature enhancement of only 17 °C would be needed relative to a bulk temperature of 150 °C. Microwave energy can provide that temperature increase instantly. Likewise, for a 100-fold rate increase, the temperature would have to reach 185 °C and would require approximately a 35 °C increase over the bulk temperature. A 1000-fold enhancement would need a 56 °C increase. These instantaneous temperatures are very consistent with the temperatures that would be expected in a microwave system and are directly responsible for the reaction rate and yield enhancements.

Figure 8

Enhanced reaction rates

$$k = Ae^{-E_a/RT}$$

For T_{bulk} *= 150 °C*

and

E_a = 50 *k*cal/mol

($T_{instantaneous} > T_{bulk}$)

1000 x rate	$T_{instantaneous}$ = 206 °C
100 x rate	$T_{instantaneous}$ = 185 °C
10 x rate	$T_{instantaneous}$ = 167 °C

IV. What types of chemical reactions would be most affected?

There are two main types of chemical reactions, **kinetic** and **thermodynamic** (Figure 9). Chemical reactions driven by conventional heating are more likely to perform under kinetic control (Reaction 1, Figure 9). These reactions usually require only mild conditions. A resonance-stabilized intermediate will take the easiest path — one with the lowest activation energy — to its products. Alternatively, thermodynamically controlled reactions have higher activation energies and require harsh conditions to complete (Reaction 2, Figure 9). In microwave driven reactions, the molecules are provided powerful instantaneous energy, which allows them to reach these higher activation energy levels and leads to the thermodynamic product. This mechanism is a probable explanation for some of the work that has

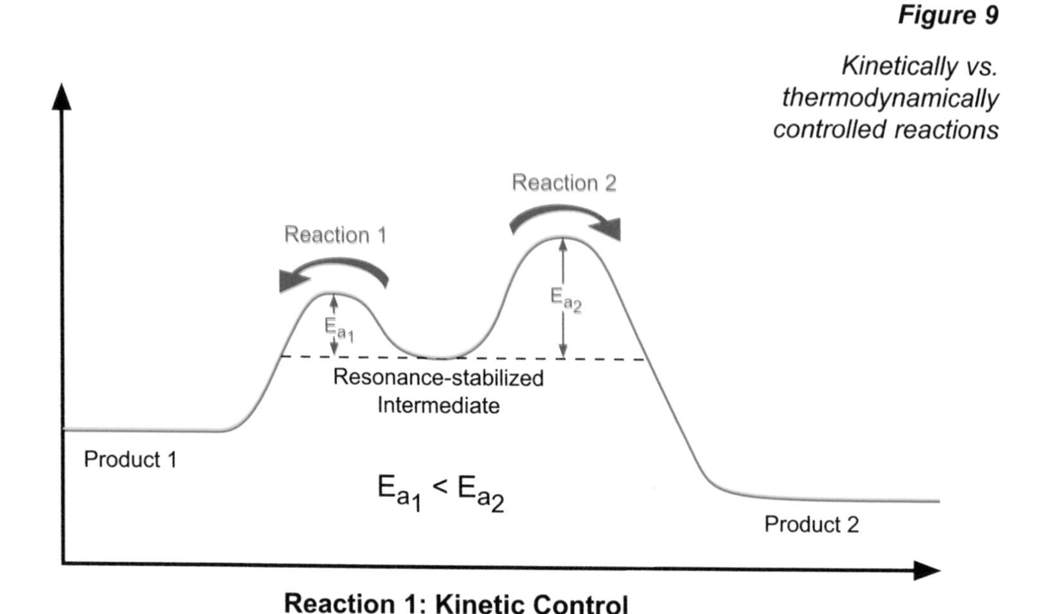

Figure 9

Kinetically vs. thermodynamically controlled reactions

been done recently on highly diastereoselective syntheses, which were generated using microwave irradiation versus conventional heating.[16]

Clearly, microwave heating is extremely useful in slower reactions where high activation energies are required to do various transformations. Empirically, the activation energy parameter expresses the temperature dependence of the rate constant. A small E_a corresponds to a rate constant that does not increase rapidly with temperature, whereas a reaction with strong temperature dependence has a large E_a. With the elevated molecular energy generated by the transfer of microwave energy, reactions that required many hours or even days to complete have been accomplished in minutes. It is also possible to use non-polar solvents to actually reduce bulk heating and directly energize the molecule. The solvent acts as a heat sink to pull thermal heat away from the reactants. The use of non-polar solvents in this manner will open opportunities to perform temperature-sensitive reactions that were not possible with conventional heating. This will be discussed in greater detail in Chapter 2. Microwave-enhanced synthesis greatly expands the options organic chemists have in their search for new compounds. Drug discovery can be taken to new heights as chemists explore the depths of their creativity.

References for Chapter 1

1. Neas, E.D.; Collins, M.J. *Introduction to Microwave Sample Preparation Theory and Practice*, Kingston, H.M.; Jassie, L.B., Eds., American Chemical Society **1988**, ch. 2, pp. 7-32.

2. Mingos, D.M.P.; Baghurst, D.R. *Microwave-Enhanced Chemistry Fundamentals, Sample Preparation, and Applications*, Kingston, H.M.; Haswell, S.J., Eds., American Chemical Society **1997**, ch. 1, pp. 3-53.

3. Giguere, R.J.; Bray, T.L.; Duncan, S.M.; Majetich, G. "Application of commercial microwave ovens to organic synthesis." *Tetrahedron Lett.* **1986**, *27*, pp. 4945-48.

4. Gedye, R.; Smith, F.; Westaway, K.; Ali, H.; Baldisera, L.; Laberge, L.; Rousell, J. "The use of microwave ovens for rapid organic synthesis." *Tetrahedron Lett.* **1986**, *27*, pp. 279-82.

5. Jun, C.H.; Chung, J.H.; Lee, D.Y.; Loupy, A.; Chatti, S. "Solvent-free chelation-assisted intermolecular hydroacylation: effect of microwave irradiation in the synthesis of ketone from aldehyde and 1-alkene by Rh(I) complex." *Tetrahedron Lett.* **2001**, *42*, pp. 4803-05.

6. Loupy, A.; Perreus, L.; Liagre, M.; Burle, K.; Moneuse, M. "Reactivity and selectivity under microwaves in organic chemistry. Relation with medium effects and reaction mechanisms." *Pure Appl. Chem.* **2001**, *73*, pp. 161-66.

7. Mingos, D.M.P.; Baghurst, D.R. "Applications of microwave dielectric heating effects to synthetic problems in chemistry." *Chem. Soc. Rev.* **1991**, *20*, pp. 1-47.

8. Loupy, A.; Petit, A.; Hamelin, J.; Texier-Boullet, F.; Jacquault, P.; Mathe, D. "New solvent-free organic synthesis using focused microwaves." *Synthesis* **1998**, 9, pp. 1213-34.

9. Loupy, A. "Microwaves in organic synthesis: a clean and high-performance methodology." *Spectra Anal.* **1993**, *22*, p. 175.

10. Majetich, G.; Hicks, R. "Applications of microwave-accelerated organic synthesis." *Radiat. Phys. Chem.* **1995**, *45*, pp. 567-79.

11. Bose, A.K.; Manhas, M.S.; Ghosh, M.; Shah, M.; Raju, V.S.; Bari, S.S.; Newaz, S.N.; Banik, B.K.; Chaudhary, A.G.; Barakat, K.J. "Microwave-induced organic reaction enhancement chemistry. 2. Simplified techniques." *J. Org. Chem.* **1991**, *56*, pp. 6968-70.

12. Johannsson, H. "A solution to the bottleneck in drug discovery." *Am. Laboratory* **2001**, *33*, pp. 28-32.

13. Larhed, M.; Hallberg, "A. Microwave-assisted high-speed chemistry: a new technique in drug discovery." *Drug Discovery Today* **2001**, *6*, pp. 406-16.

14. Strauss, C.R.; Trainor, R.W. "Developments in microwave-assisted organic chemistry." *Aust. J. Chem.* **1995**, *48*, pp. 1665-92.

15. Langa, F.; De La Cruz, P.; De La Hoz, A.; Diaz-Ortiz, A.; Diez-Barra, E. "Microwave irradiation: more than just a method for accelerating reactions." *Contemp. Org. Synth.* **1997**, *4*, pp. 373-86.

16. Kuhnert, N.; Danks, T.N. "Highly diastereoselective synthesis of 1,3-ozazolidines under thermodynamic control using focused microwave irradiation under solvent-free conditions." *Green Chem.* **2001**, *3*, pp. 68-70.

Chapter 2
Solvents

Solvents play a very important role in organic synthesis. Most reactions take place in solution, and therefore, choice of solvent can be a crucial factor in the outcome of a reaction. One of the most important characteristics of a solvent is its polarity. With microwave heating, this becomes a more significant component, as microwaves directly couple with the molecules that are present in the reaction mixture. The more polar a reaction mixture is, the greater its ability to couple with the microwave energy. As discussed in the previous chapter, this interaction leads to a rapid rise in temperature and faster reaction rates. This chapter will discuss the theory behind solvent polarity and how it pertains to the individual solvents, their physical constants, and how they behave in a microwave field. In addition, the last section will discuss how to choose a solvent in a microwave-enhanced organic reaction.

The more efficient a solvent is in coupling with the microwave energy, the faster the temperature of the reaction mixture increases.

I. Theory

Many factors characterize the polarity of a solvent. Intrinsically, the dielectric constant, dipole moment, dielectric loss, tangent delta, and dielectric relaxation time all contribute to an individual solvent's **absorbing characteristics**. The **dielectric constant (ε')**, also known as the relative permittivity, of a solvent measures its ability to store electric charges. Mathematically, it is the ratio of the electrical capacity of a capacitor filled with the solvent to the electrical capacity of the evacuated capacitor ($\varepsilon'=C_{filled}/C_{evacuated}$). This value, when measured, is dependent on both temperature and frequency.

The **dipole moment**, which is measured in Debye units (D), is also a mathematical entity. It is the product of the distance between the centers of charge in the solvent molecule multiplied by the magnitude of that charge. One equation used to determine dipole moment is: $T = pE$ (T = torque, p = dipole moment, and E = field strength). The magnitude can also be defined as: $\mu = Qr$ (μ = dipole moment, Q = charge, and r = distance between charges). Molecules with large dipole moments also have large dielectric constants. This is because polarization depends on **dipole rotation** — the ability of a molecule's dipole to align with a rapidly changing electric field.[17]

The ability of a substance to convert electromagnetic energy into heat at a given frequency and temperature is determined by the following equation: $\tan \delta = \varepsilon''/\varepsilon'$. **Tangent delta ($\delta$)**, or loss tangent, is the dissipation factor of the sample or how efficiently microwave energy is converted into thermal energy. It is defined as the ratio of the dielectric loss, or complexed permittivity (ε''), to the dielectric constant (ε'). **Dielectric loss** is the amount of input microwave energy that is lost to the sample by being

Loss Tangent Equation

$\tan \delta = \varepsilon''/\varepsilon'$

dissipated as heat. It is this value, ε″, that best provides the organic chemist with the coupling efficiency of a particular solvent. This will be discussed in greater detail in the following section.

The three main dielectric parameters, tangent delta, dielectric constant, and dielectric loss, are all related to the ability of a solvent to absorb microwave energy. Molecular relaxation time has a large effect on these parameters. The dielectric relaxation time is the time it takes a molecule to achieve 63% of its return to randomized disorder from an organized state after an applied microwave field is removed.[1] Functional groups, temperature, frequency, and volume will all influence the relaxation time of a solvent. Most commercial microwave systems are set to a frequency of 2450 MHz.

Figure 10

Tangent Delta versus Temperature

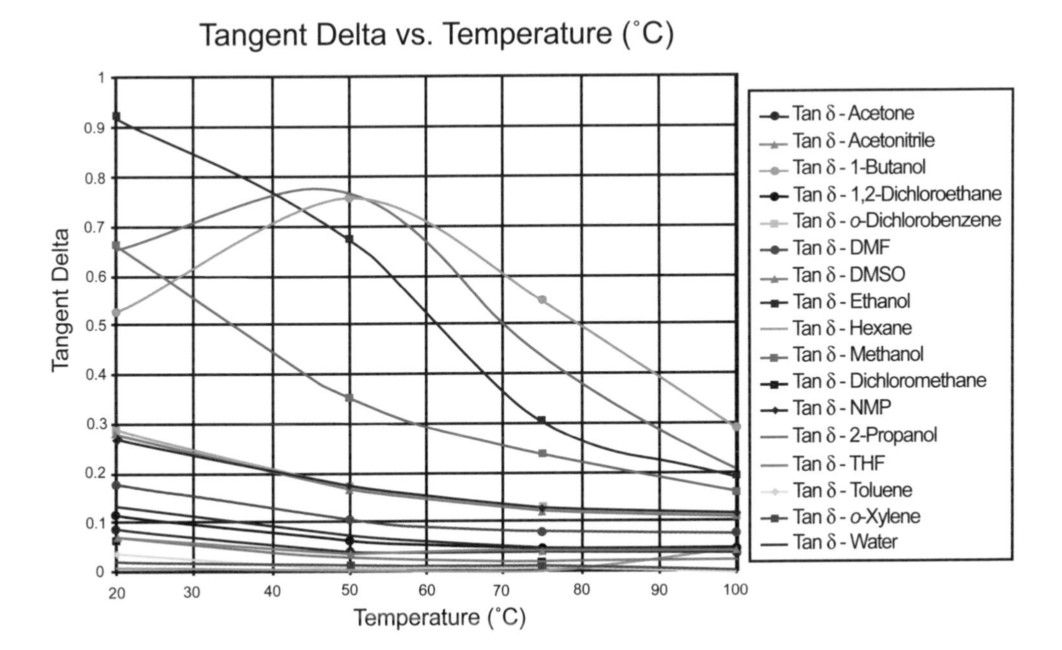

Figure 11

Dielectric Constant versus Temperature

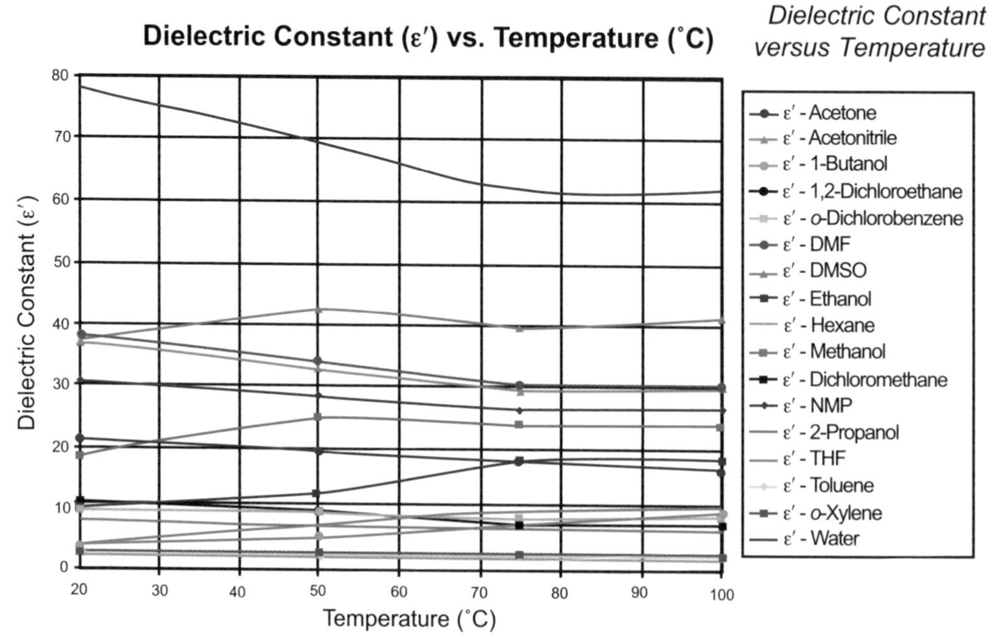

Figure 12

Dielectric Loss versus Temperature

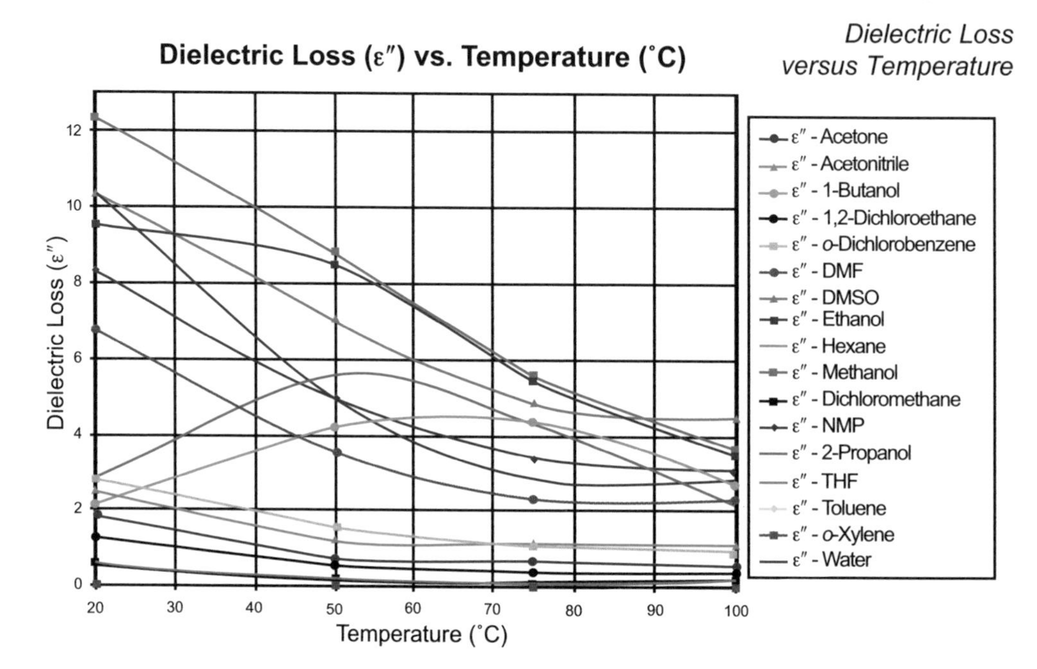

At this frequency, the only thing that can change the three parameters is temperature. As the temperature of a solvent increases, a decrease in its relaxation time and dielectric parameters will be seen, and hence, its coupling efficiency. There are a few exceptions, but this is generally the trend. A graphical representation of this effect for the tangent delta, dielectric constant, and dielectric loss values of 17 common solvents is shown in Figures 10-12, respectively.

II. Solvents

When considering solvents for a microwave-enhanced organic reaction, a chemist must now realize that boiling points become a less important factor in that decision. Microwave energy (300 W) will reach and bypass the boiling point of most solvents in a matter of seconds. Using pressurized reaction vessels provide for greater use of the lower boiling point solvents that are normally ignored in conventional high temperature reactions. Alternatively, one factor that becomes more important is how efficient the molecules in a solvent or a solvent mixture couple with an applied microwave field.

Microwave energy will reach and bypass the boiling point of most solvents in a matter of seconds.

As we established in the previous section, the three main dielectric parameters all factor into the ability of a solvent to absorb microwave energy, but they do so quite differently. Table 1 has been developed to show this difference in thirty common solvents. There are three main columns (dielectric constant, tan δ, and dielectric loss, respectively), which are indicated by the bold lines. The data, which was measured at room temperature and at a frequency of 2450 MHz, is shown in descending order.[17,18a] We find that the values, and their respective solvents, represented in the third column (dielectric loss) are most indicative of how quickly a solvent will reach its desired temperature. In general, the higher the number, the more efficient the solvent converts microwave energy into thermal energy, and thus, the faster the temperature will increase.

The solvents in Table 1 can easily be categorized into three different groups: high, medium, and low absorbing solvents. By examining the dielectric loss values from the third column of the table, one can see where significant gaps in between the numbers are present (bold lines). The high absorbing solvents are ones that have dielectric losses greater than 14.00. Medium absorbers would generally have dielectric loss values between 1.00 and 13.99, and low absorbing molecules have dielectric losses that are less than 1.00. High absorbers like small chain alcohols, dimethyl sulfoxide (DMSO), and nitrobenzene all have large dielectric losses, so they heat very quickly within the microwave chamber. Common organic solvents that are grouped as medium absorbers include dimethylformamide (DMF), acetonitrile, butanols, ketones, and water. These, too, heat very efficiently, but they require more time to reach desired temperatures. Additionally, chloroform, dichloromethane, ethyl acetate, and, as expected, ethers and hydrocarbons, are very low microwave absorbing solvents. They can be heated to temperatures well above their boiling points, but they take much longer.

Water, for example, has the highest dielectric constant (80.4) of the thirty solvents, but its tangent delta and dielectric loss values do not rank at the top of their respective lists. If we only considered the dielectric constant, we would assume that water is the most polar solvent in a microwave field. This is not the case. It should be classified as a medium absorber, which is where the second and third columns (tan δ and dielectric loss) categorize it. In another example, acetonitrile is ranked fairly high in the dielectric constant column with a value of 37.5. Looking at the tangent delta values, acetonitrile plummets close to the bottom at 0.062. So, what kind of

Table 1

Dielectric constant (ε′), tan δ, and dielectric loss (ε″) for 30 common solvents (measured at room temperature and 2450 MHz)

Solvent (bp °C)	Dielectric Constant (ε′)	Solvent	Tan δ	Solvent	Dielectric Loss (ε″)
Water (100)	80.4	Ethylene Glycol	1.350	Ethylene Glycol	49.950
Formic Acid (100)	58.5	Ethanol	.941	Formic Acid	42.237
DMSO (189)	45.0	DMSO	.825	DMSO	37.125
DMF (153)	37.7	2-Propanol	.799	Ethanol	22.866
Acetonitrile (82)	37.5	1-Propanol	.757	Methanol	21.483
Ethylene Glycol (197)	37.0	Formic Acid	.722	Nitrobenzene	20.497
Nitromethane (101)	36.0	Methanol	.659	1-Propanol	15.216
Nitrobenzene (202)	34.8	Nitrobenzene	.589	2-Propanol	14.622
Methanol (65)	32.6	1-Butanol	.571	Water	9.889
NMP (215)	32.2	Isobutanol	.522	1-Butanol	9.764
Ethanol (78)	24.3	2-Butanol	.447	NMP	8.855
Acetone (56)	20.7	2-Methoxyethanol	.410	Isobutanol	8.248
1-Propanol (97)	20.1	*o*-Dichlorobenzene	.280	2-Butanol	7.063
MEK (80)	18.5	NMP	.275	2-Methoxyethanol	6.929
2-Propanol (82)	18.3	Acetic Acid	.174	DMF	6.070
1-Butanol (118)	17.1	DMF	.161	*o*-Dichlorobenzene	2.772
2-Methoxyethanol (124)	16.9	1,2-Dichloroethane	.127	Acetonitrile	2.325
2-Butanol (100)	15.8	Water	.123	Nitromethane	2.304
Isobutanol (108)	15.8	Chlorobenzene	.101	MEK	1.462
1,2-Dichloroethane (83)	10.4	Chloroform	.091	1,2-Dichloroethane	1.321
o-Dichlorobenzene (180)	9.9	MEK	.079	Acetone	1.118
Dichloromethane (40)	9.1	Nitromethane	.064	Acetic Acid	1.079
THF (66)	7.4	Acetonitrile	.062	Chloroform	0.437
Acetic Acid (113)	6.2	Ethyl Acetate	.059	Dichloromethane	0.382
Ethyl Acetate (77)	6.0	Acetone	.054	Ethyl Acetate	0.354
Chloroform (61)	4.8	THF	.047	THF	0.348
Chlorobenzene (132)	2.6	Dichloromethane	.042	Chlorobenzene	0.263
o-Xylene (144)	2.6	Toluene	.040	Toluene	0.096
Toluene (111)	2.4	Hexane	.020	*o*-Xylene	0.047
Hexane (69)	1.9	*o*-Xylene	.018	Hexane	0.038

solvent is acetonitrile? Once again, we look to the third column where acetonitrile falls in the middle with a dielectric loss value of 2.325. Acetonitrile should be considered as a medium absorber.

In addition to the coupling efficiency of a solvent, a chemist should also be aware of the pressures that are generated at certain temperatures in a sealed tube for that solvent. A pressurized environment can be very advantageous to many different kinds of chemistries. As the temperature of a solvent increases above its boiling point, more and more pressure builds up in the reaction vessel. In solvent-only experiments, the pressure that is generated for a specific temperature is independent of the solvent volume to head space ratio. (The head space is the volume of a 10-mL capacity pressure tube that is not occupied by the solvent at room temperature.) This will not be true for actual chemical reactions, as newly formed molecules are constantly being introduced into the gas phase throughout the duration of the reaction. Tables 2-26 show the pressures generated at specific temperatures for 25 common solvents at 1-, 3-, and 5-mL volumes (with head space volumes of 9-, 7-, and 5-mL volumes, respectively), using 300 W of microwave power. Accompanying each table is a figure (Figures 13-37) that shows the temperature and pressure curves for 3-mL of that particular solvent in a 10-mL capacity pressure tube. These solvents were subjected to a rigorous method of 300 W, 250 °C, and a 300 psi limit for ten minutes. Upon reaching 250 °C, the instrument held for 10 seconds and then cooled. Some solvents never reached 250 °C in the ten minute period, and this can be seen in the figures with longer run times. For most microwave synthesis instruments if the pressure inside the vessel reaches 300 psi, the instrument will abort its run (see dichloromethane, ethanol, and methanol). The figures display the approximate maximum pressure and temperature that each solvent can achieve.

Table 2

Pressures generated at specific temperatures of acetone for different volume to head space ratios

Solvent (bp) and Volume	Temperature (°C) /Pressure (psi)	BP + 10°	BP + 25°	BP + 50°
Acetone (56)				
1 mL	Temperature	66	81	106
	Pressure	4	13	40
3 mL	Temperature	66	81	106
	Pressure	4	12	40
5 mL	Temperature	66	81	106
	Pressure	4	8	40

Figure 13

Temperature and pressure curves for 3 mL of acetone

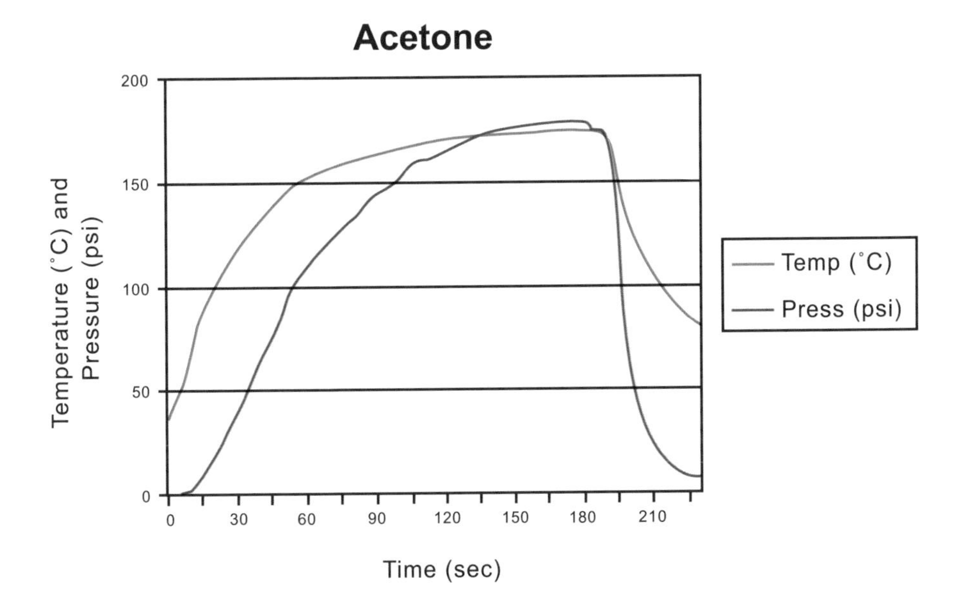

Table 3

Pressures generated at specific temperatures of acetonitrile for different volume to head space ratios

Solvent (bp) and Volume	Temperature (°C) /Pressure (psi)	BP + 10°	BP + 25°	BP + 50°
Acetonitrile (82)				
1 mL	Temperature	92	107	132
	Pressure	4	10	35
3 mL	Temperature	92	107	132
	Pressure	4	11	38
5 mL	Temperature	92	107	132
	Pressure	6	11	35

Figure 14

Temperature and pressure curves for 3 mL of acetonitrile

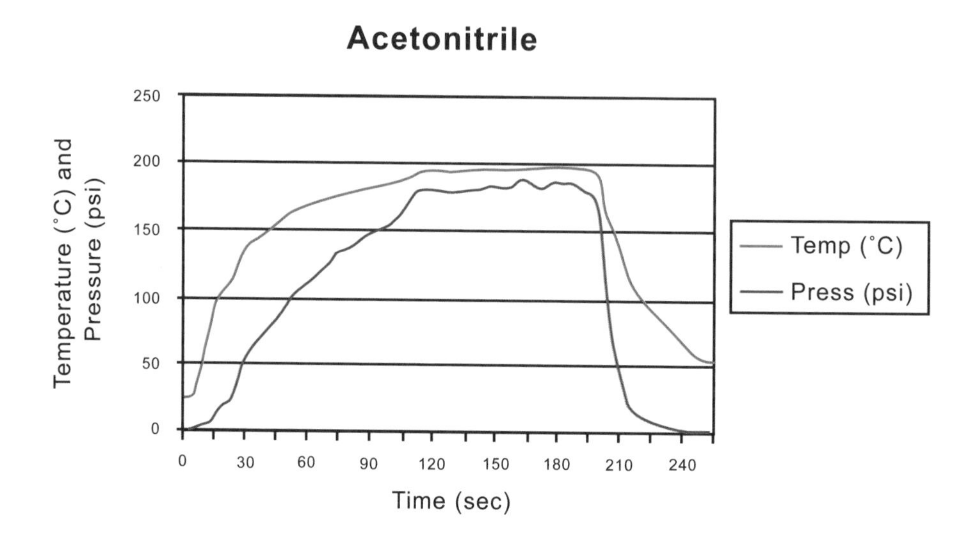

Table 4

Pressures generated at specific temperatures of 1-butanol for different volume to head space ratios

Solvent (bp) and Volume	Temperature (°C) /Pressure (psi)	BP + 10°	BP + 25°	BP + 50°
1-Butanol (118)				
1 mL	Temperature	128	143	168
	Pressure	7	13	41
3 mL	Temperature	128	143	168
	Pressure	13	20	41
5 mL	Temperature	128	143	168
	Pressure	15	22	45

Figure 15

Temperature and pressure curves for 3 mL of 1-butanol

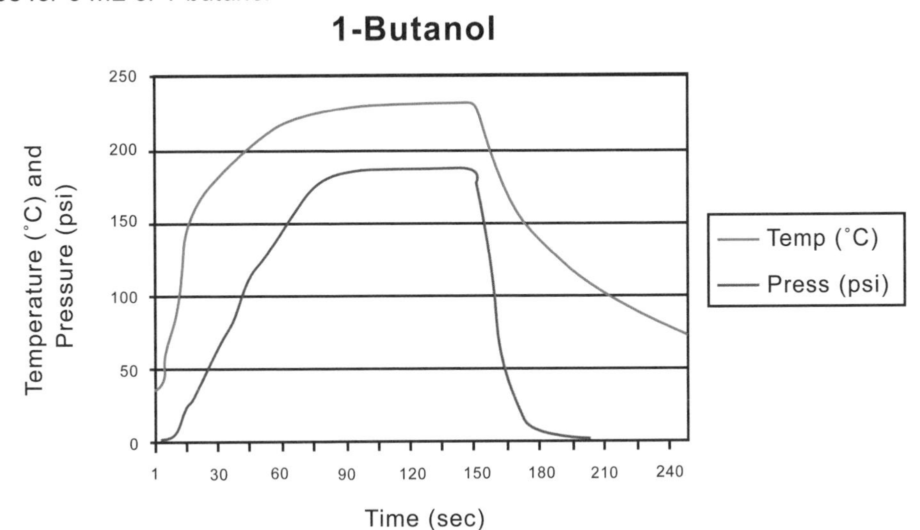

Table 5

Pressures generated at specific temperatures of chloroform for different volume to head space ratios

Solvent (bp) and Volume	Temperature (°C) /Pressure (psi)	BP + 10°	BP + 25°	BP + 50°
Chloroform (61)				
1 mL	Temperature	71	86	111
	Pressure	9	16	44
3 mL	Temperature	71	86	111
	Pressure	8	15	40
5 mL	Temperature	71	86	111
	Pressure	9	14	42

Figure 16

Temperature and pressure curves for 3 mL of chloroform

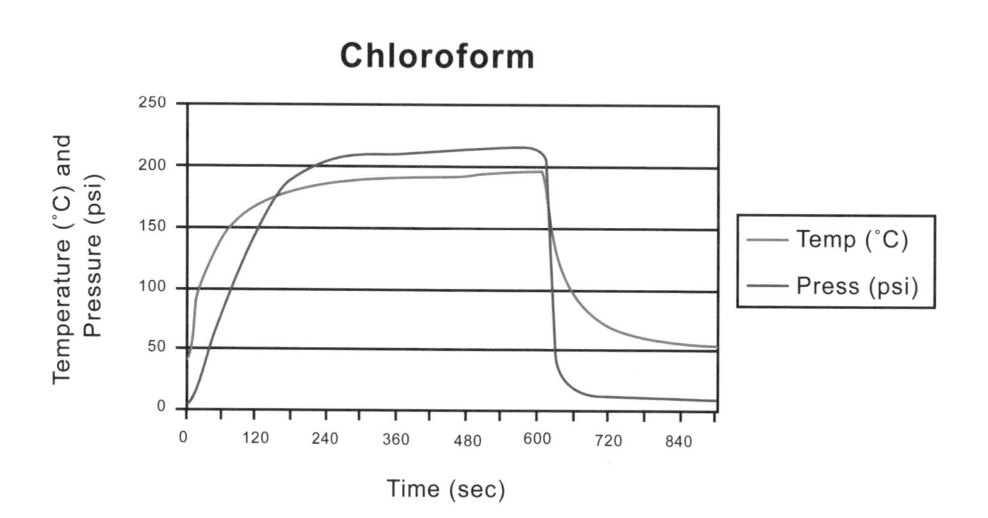

Table 6

Pressures generated at specific temperatures of o-dichlorobenzene for different volume to head space ratios

Solvent (bp) and Volume	Temperature (°C) /Pressure (psi)	BP + 10°	BP + 25°	BP + 50°
o-Dichlorobenzene (180)				
1 mL	Temperature	190	205	230
	Pressure	4	12	30
3 mL	Temperature	190	205	230
	Pressure	6	11	32
5 mL	Temperature	190	205	230
	Pressure	7	11	30

Figure 17

Temperature and pressure curves for 3 mL of o-dichlorobenzene

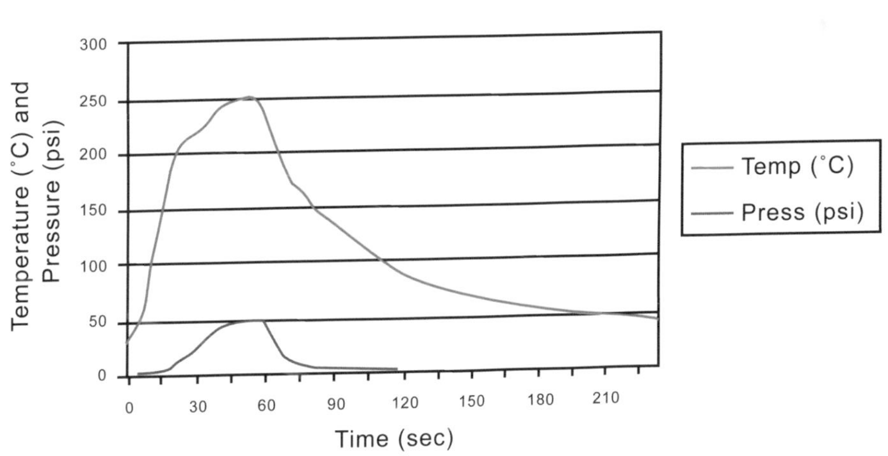

Table 7

Pressures generated at specific temperatures of 1,2-dichloroethane for different volume to head space ratios

Solvent (bp) and Volume	Temperature (°C) /Pressure (psi)	BP + 10°	BP + 25°	BP + 50°
1,2-Dichloroethane (83)				
1 mL	Temperature	93	108	133
	Pressure	4	10	35
3 mL	Temperature	93	108	133
	Pressure	3	10	35
5 mL	Temperature	93	108	133
	Pressure	4	10	34

Figure 18

Temperature and pressure curves for 3 mL of 1,2-dichloroethane

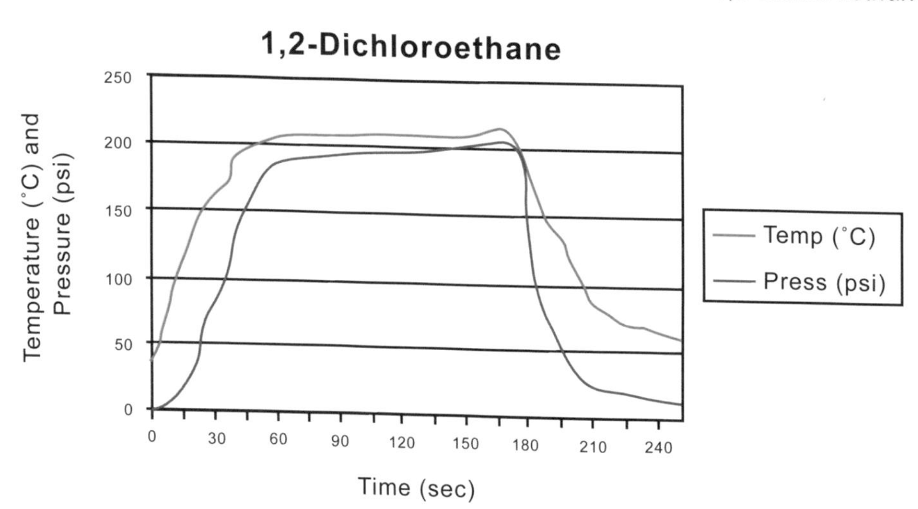

Table 8

Pressures generated at specific temperatures of dichloromethane for different volume to head space ratios

Solvent (bp) and Volume	Temperature (°C) /Pressure (psi)	BP + 10°	BP + 25°	BP + 50°
Dichloromethane (40)				
1 mL	Temperature	50	65	90
	Pressure	4	10	32
3 mL	Temperature	50	65	90
	Pressure	4	10	37
5 mL	Temperature	50	65	90
	Pressure	2	8	42

Figure 19

Temperature and pressure curves for 3 mL of dichloromethane

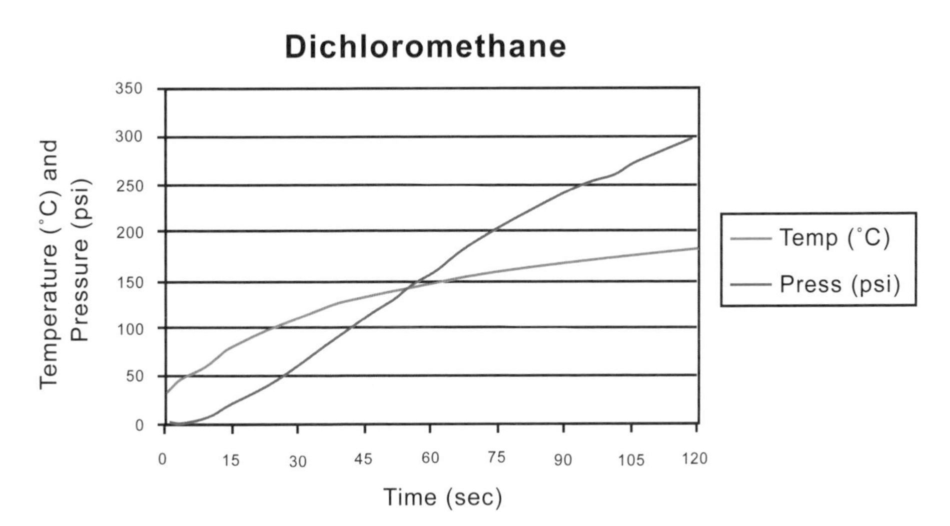

Table 9

Pressures generated at specific temperatures of DMF for different volume to head space ratios

Solvent (bp) and Volume	Temperature (°C) /Pressure (psi)	BP + 10°	BP + 25°	BP + 50°
DMF (153)				
1 mL	Temperature	163	178	203
	Pressure	6	13	35
3 mL	Temperature	163	178	203
	Pressure	10	16	38
5 mL	Temperature	163	178	203
	Pressure	7	16	37

Figure 20

Temperature and pressure curves for 3 mL of DMF

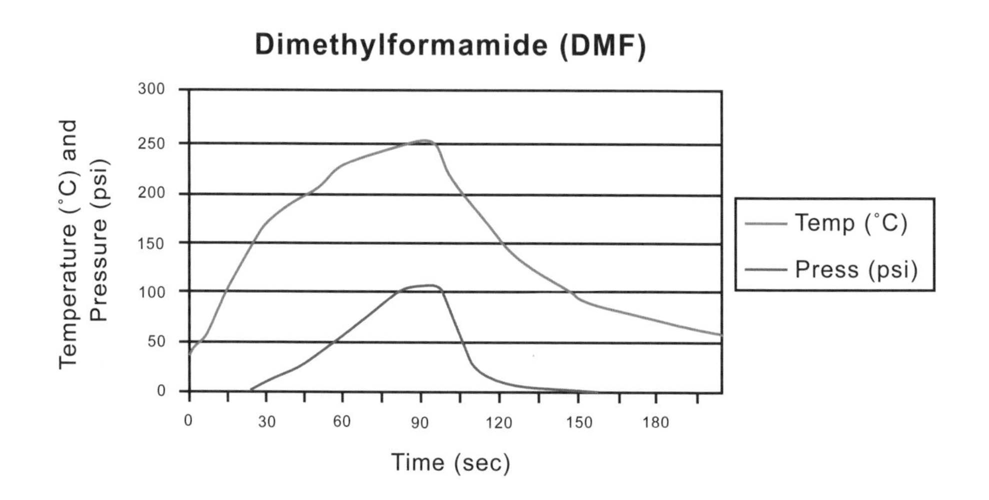

Table 10

Pressures generated at specific temperatures of DMSO for different volume to head space ratios

Solvent (bp) and Volume	Temperature (°C) /Pressure (psi)	BP + 10°	BP + 25°	BP + 50°
DMSO (189)				
1 mL	Temperature	199	214	239
	Pressure	6	14	36
3 mL	Temperature	199	214	239
	Pressure	6	14	34
5 mL	Temperature	199	214	239
	Pressure	10	17	37

Figure 21

Temperature and pressure curves for 3 mL of DMSO

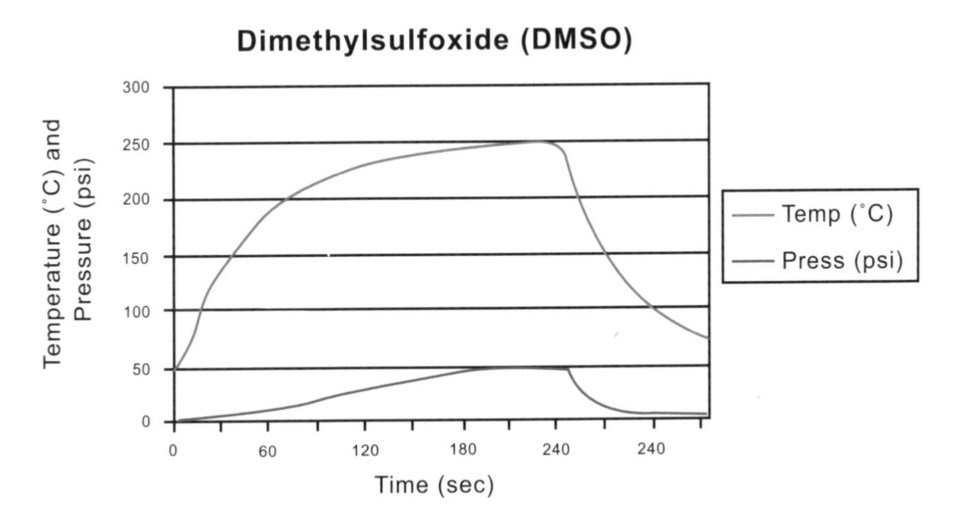

Table 11

Pressures generated at specific temperatures of 1,4-dioxane for different volume to head space ratios

Solvent (bp) and Volume	Temperature (°C) /Pressure (psi)	BP + 10°	BP + 25°	BP + 50°
1,4-Dioxane (101)				
1 mL	Temperature	111	126	203
	Pressure	10	14	30
3 mL	Temperature	111	126	203
	Pressure	7	14	22
5 mL	Temperature	111	126	203
	Pressure	7	11	19

Figure 22

Temperature and pressure curves for 3 mL of 1,4-dioxane

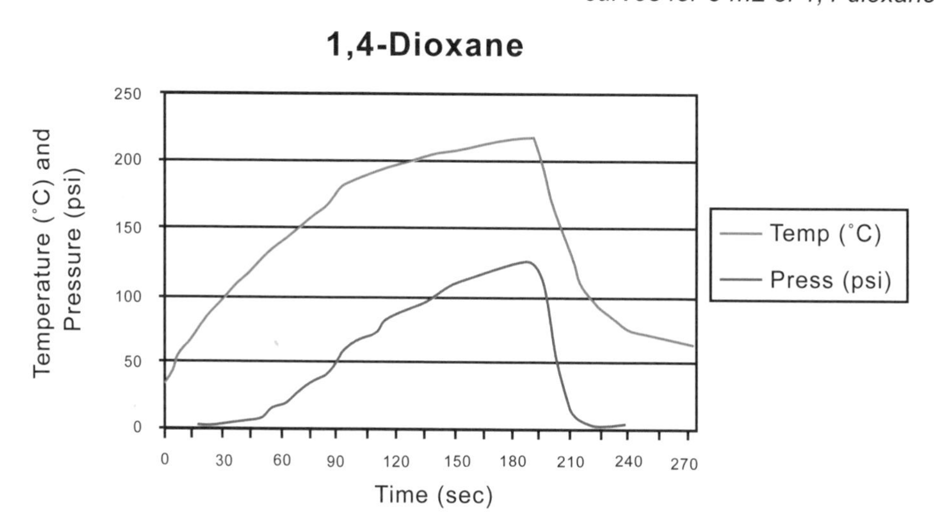

Table 12

Pressures generated at specific temperatures of ethanol for different volume to head space ratios

Solvent (bp) and Volume	Temperature (°C) /Pressure (psi)	BP + 10°	BP + 25°	BP + 50°
Ethanol (78)				
1 mL	Temperature	88	103	128
	Pressure	18	23	56
3 mL	Temperature	88	103	128
	Pressure	30	43	73
5 mL	Temperature	88	103	128
	Pressure	18	33	63

Figure 23

Temperature and pressure curves for 3 mL of ethanol

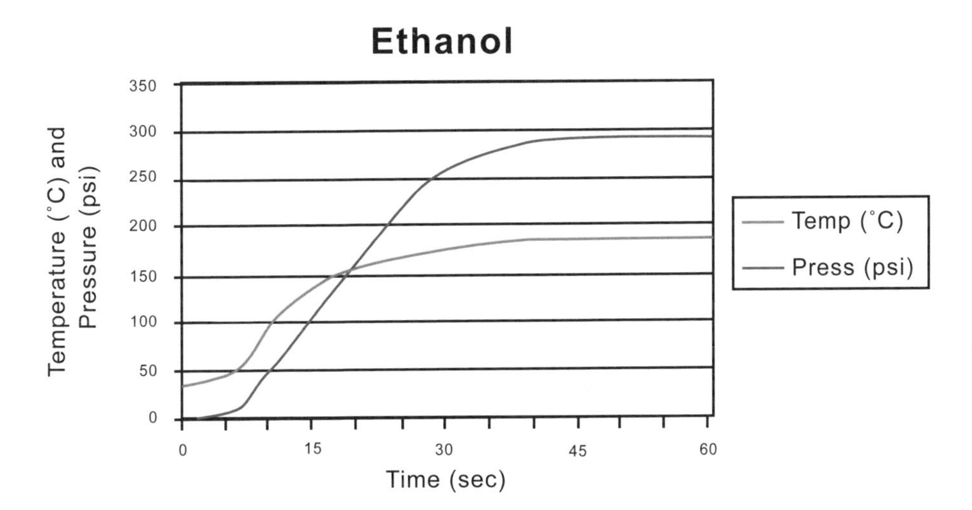

Table 13

Pressures generated at specific temperatures of ethyl acetate for different volume to head space ratios

Solvent (bp) and Volume	Temperature (°C) /Pressure (psi)	BP + 10°	BP + 25°	BP + 50°
Ethyl Acetate (77)				
1 mL	Temperature	87	102	127
	Pressure	8	13	36
3 mL	Temperature	87	102	127
	Pressure	8	14	39
5 mL	Temperature	87	102	127
	Pressure	9	12	39

Figure 24

Temperature and pressure curves for 3 mL of ethyl acetate

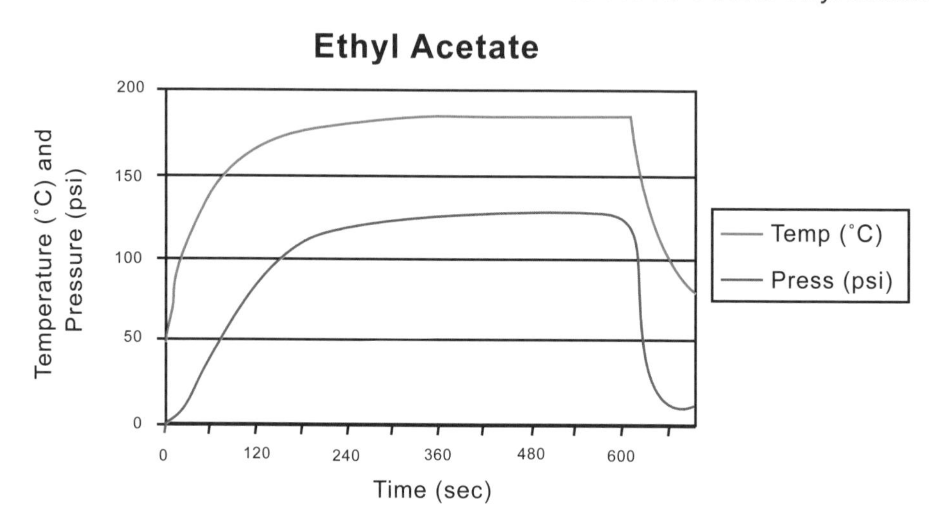

Table 14

Pressures generated at specific temperatures of HMPA for different volume to head space ratios

Solvent (bp) and Volume	Temperature (°C) /Pressure (psi)	BP + 10°	BP + 25°	BP + 50°
HMPA (231)				
1 mL	Temperature	241	-	-
	Pressure	8	-	-
3 mL	Temperature	241	-	-
	Pressure	8	-	-
5 mL	Temperature	241	-	-
	Pressure	10	-	-

Figure 25

Temperature and pressure curves for 3 mL of HMPA

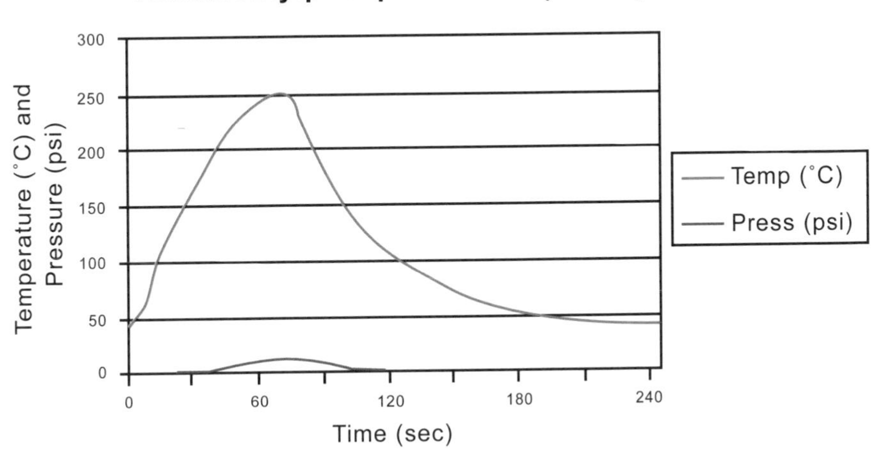

Table 15

Pressures generated at specific temperatures of hexane for different volume to head space ratios

Solvent (bp) and Volume	Temperature (°C) /Pressure (psi)	BP + 10°	BP + 25°	BP + 50°
Hexane (69)				
1 mL	Temperature	79	94	119
	Pressure	3	10	26
3 mL	Temperature	79	94	119
	Pressure	3	10	23
5 mL	Temperature	79	94	119
	Pressure	4	8	20

Figure 26

Temperature and pressure curves for 3 mL of hexane

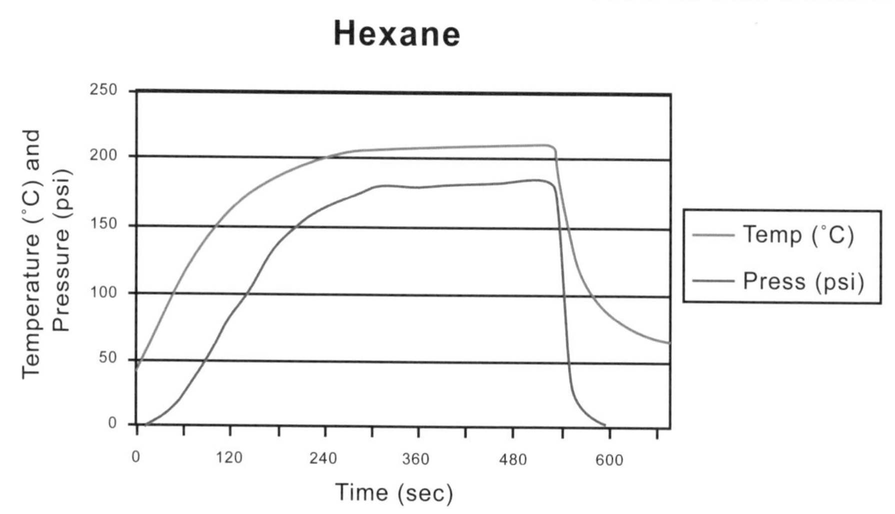

Table 16

Pressures generated at specific temperatures of methanol for different volume to head space ratios

Solvent (bp) and Volume	Temperature (°C) /Pressure (psi)	BP + 10°	BP + 25°	BP + 50°
Methanol (65)				
1 mL	Temperature	75	90	115
	Pressure	9	17	50
3 mL	Temperature	75	90	115
	Pressure	9	16	48
5 mL	Temperature	75	90	115
	Pressure	9	19	56

Figure 27

Temperature and pressure curves for 3 mL of methanol

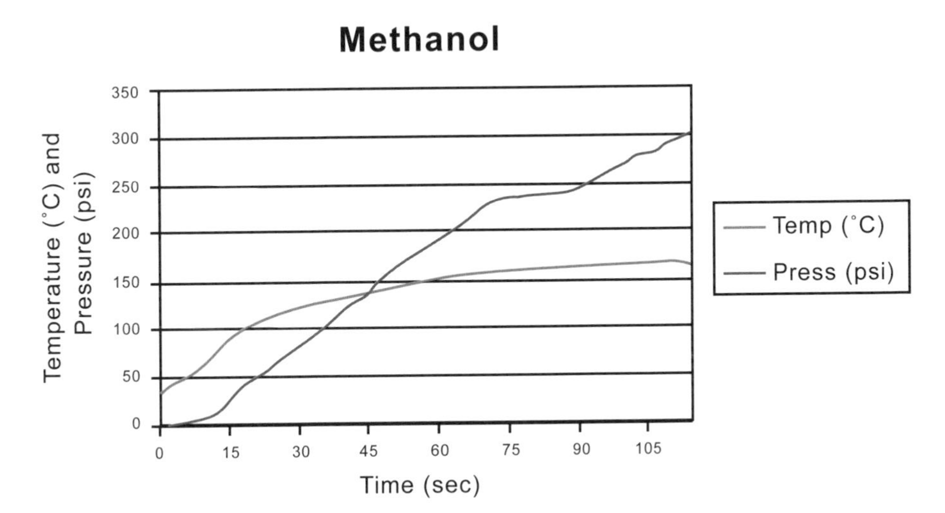

Table 17

Pressures generated at specific temperatures of methyl t*-butyl ether for different volume to head space ratios*

Solvent (bp) and Volume	Temperature (°C) /Pressure (psi)	BP + 10°	BP + 25°	BP + 50°
MTBE (55)				
1 mL	Temperature	65	80	105
	Pressure	9	16	40
3 mL	Temperature	65	80	105
	Pressure	9	13	36
5 mL	Temperature	65	80	105
	Pressure	9	16	40

Figure 28

Temperature and pressure curves for 3 mL of MTBE

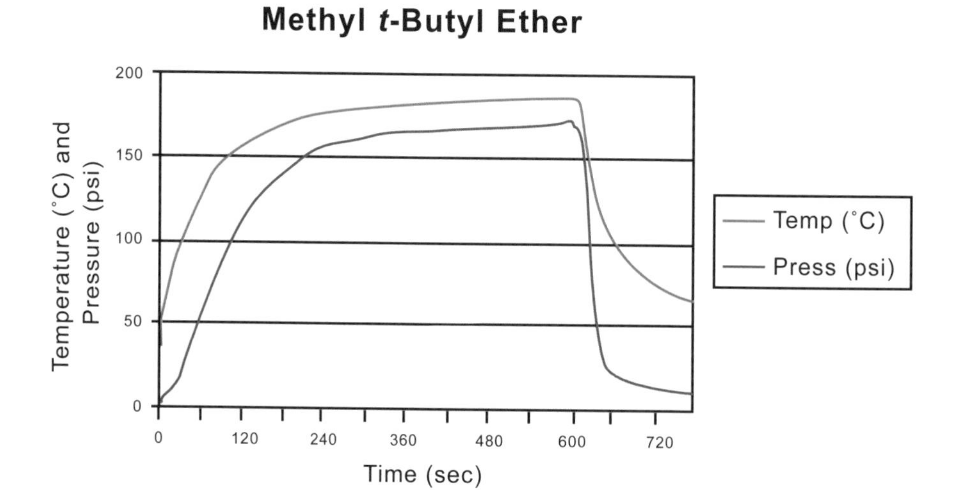

Table 18

Pressures generated at specific temperatures of methyl ethyl ketone for different volume to head space ratios

Solvent (bp) and Volume	Temperature (°C) /Pressure (psi)	BP + 10°	BP + 25°	BP + 50°
MEK (80)				
1 mL	Temperature	90	105	130
	Pressure	7	12	38
3 mL	Temperature	90	105	130
	Pressure	8	11	37
5 mL	Temperature	90	105	130
	Pressure	9	13	38

Figure 29

Temperature and pressure curves for 3 mL of MEK

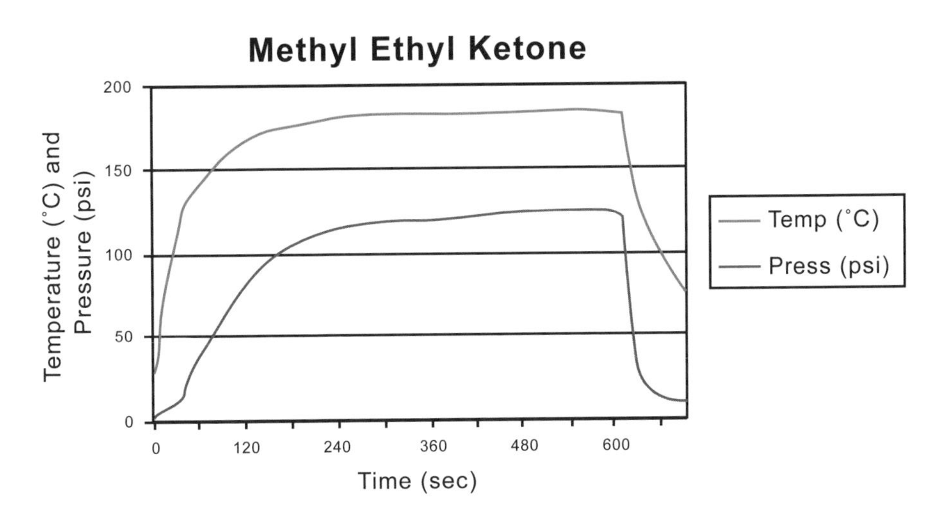

Table 19

Pressures generated at specific temperatures of NMP for different volume to head space ratios

Solvent (bp) and Volume	Temperature (°C) /Pressure (psi)	BP + 10°	BP + 25°	BP + 50°
NMP (215)				
1 mL	Temperature	225	-	-
	Pressure	8	-	-
3 mL	Temperature	225	-	-
	Pressure	8	-	-
5 mL	Temperature	225	-	-
	Pressure	11	-	-

Figure 30

Temperature and pressure curves for 3 mL of NMP

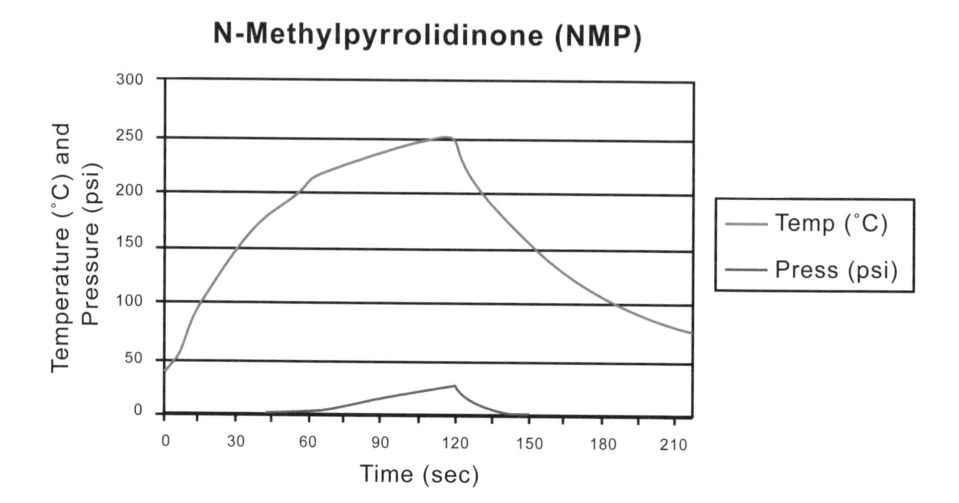

Table 20

Pressures generated at specific temperatures of nitrobenzene for different volume to head space ratios

Solvent (bp) and Volume	Temperature (°C) /Pressure (psi)	BP + 10°	BP + 25°	BP + 50°
Nitrobenzene (202)				
1 mL	Temperature	212	227	-
	Pressure	5	10	-
3 mL	Temperature	212	227	-
	Pressure	12	15	-
5 mL	Temperature	212	227	-
	Pressure	8	12	-

Figure 31

Temperature and pressure curves for 3 mL of nitrobenzene

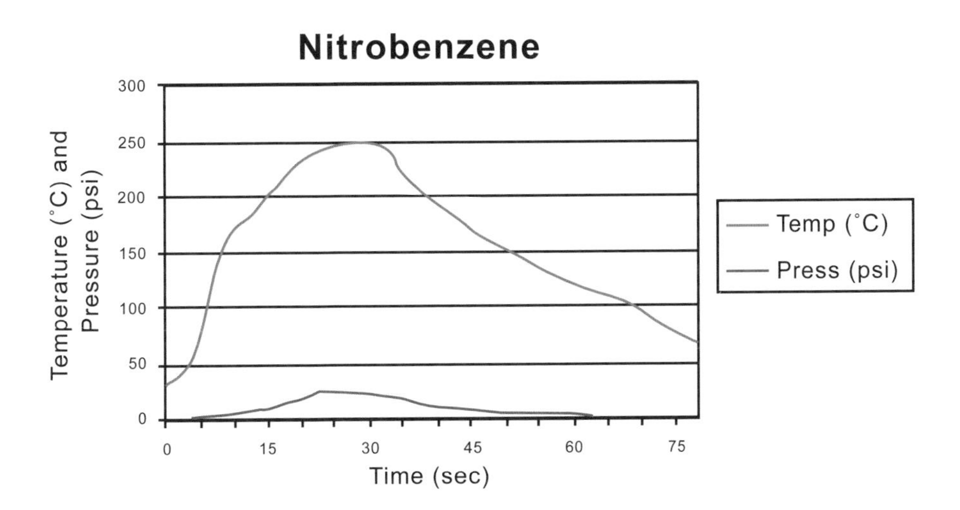

Table 21

Pressures generated at specific temperatures of pyridine for different volume to head space ratios

Solvent (bp) and Volume	Temperature (°C) /Pressure (psi)	BP + 10°	BP + 25°	BP + 50°
Pyridine (115)				
1 mL	Temperature	125	140	165
	Pressure	8	14	37
3 mL	Temperature	125	140	165
	Pressure	8	14	42
5 mL	Temperature	125	140	165
	Pressure	9	16	44

Figure 32

Temperature and pressure curves for 3 mL of pyridine

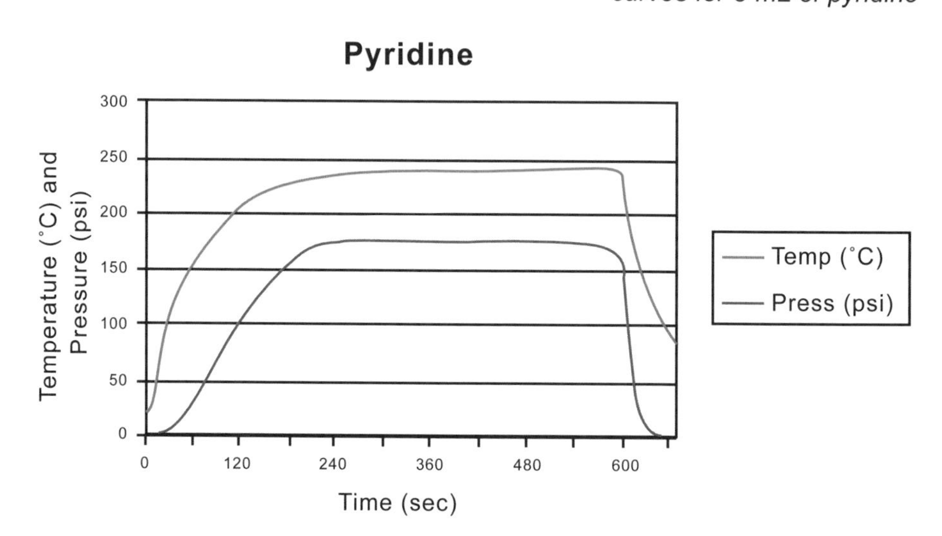

Table 22

Pressures generated at specific temperatures of THF for different volume to head space ratios

Solvent (bp) and Volume	Temperature (°C) /Pressure (psi)	BP + 10°	BP + 25°	BP + 50°
THF (66)				
1 mL	Temperature	76	91	116
	Pressure	6	12	36
3 mL	Temperature	76	91	116
	Pressure	5	10	36
5 mL	Temperature	76	91	116
	Pressure	5	12	37

Figure 33

Temperature and pressure curves for 3 mL of THF

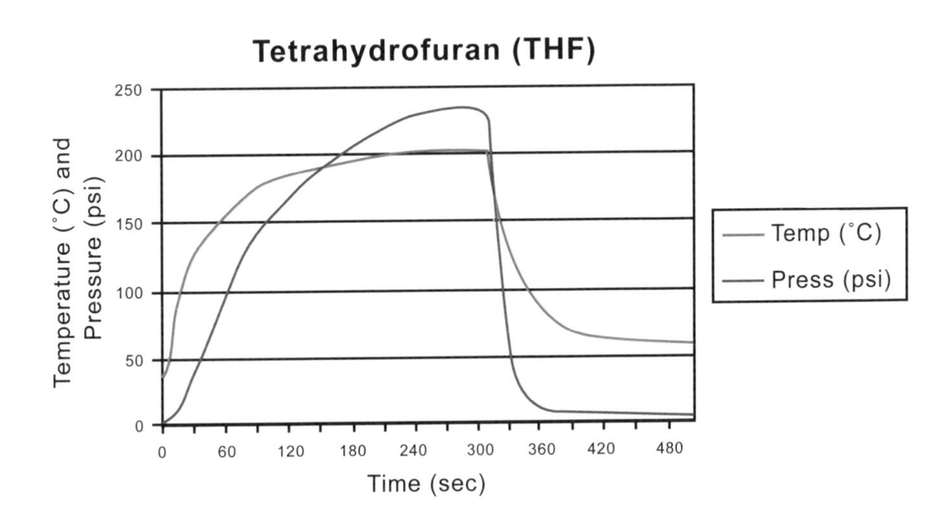

Table 23

Pressures generated at specific temperatures of toluene for different volume to head space ratios

Solvent (bp) and Volume	Temperature (°C) /Pressure (psi)	BP + 10°	BP + 25°	BP + 50°
Toluene (111)				
1 mL	Temperature	121	136	161
	Pressure	8	10	24
3 mL	Temperature	121	136	161
	Pressure	6	10	21
5 mL	Temperature	121	136	161
	Pressure	6	9	14

Figure 34

Temperature and pressure curves for 3 mL of toluene

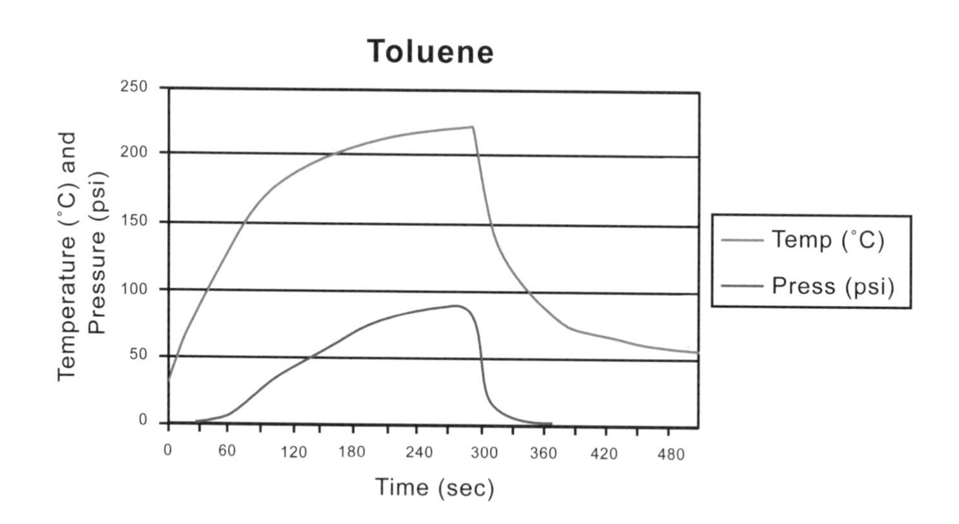

Table 24

Pressures generated at specific temperatures of triethylamine for different volume to head space ratios

Solvent (bp) and Volume	Temperature (°C) /Pressure (psi)	BP + 10°	BP + 25°	BP + 50°
Triethylamine (89)				
1 mL	Temperature	99	114	139
	Pressure	2	10	30
3 mL	Temperature	99	114	139
	Pressure	2	12	32
5 mL	Temperature	99	114	139
	Pressure	2	10	28

Figure 35

Temperature and pressure curves for 3 mL of triethylamine

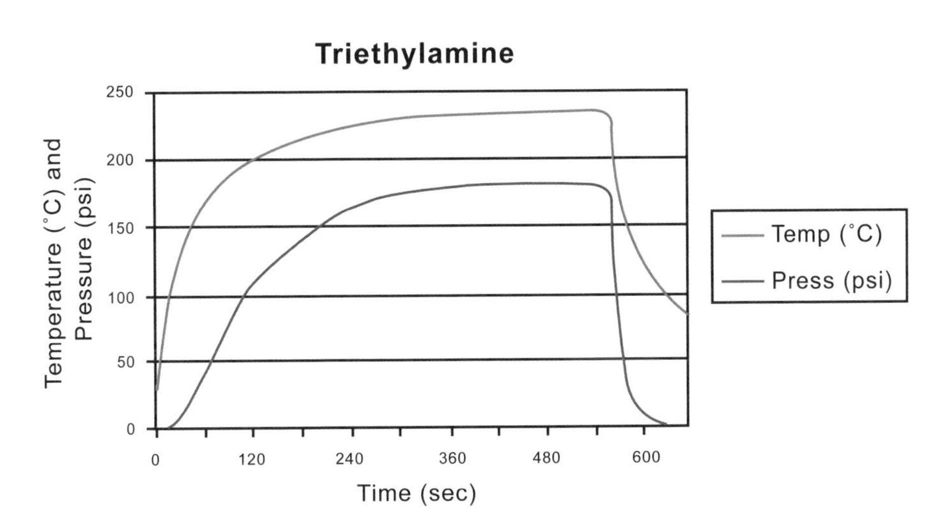

Table 25

Pressures generated at specific temperatures of water for different volume to head space ratios

Solvent (bp) and Volume	Temperature (°C) /Pressure (psi)	BP + 10°	BP + 25°	BP + 50°
Water (100)				
1 mL	Temperature	110	125	150
	Pressure	5	16	46
3 mL	Temperature	110	125	150
	Pressure	8	17	46
5 mL	Temperature	110	125	150
	Pressure	12	20	46

Figure 36

Temperature and pressure curves for 3 mL of water

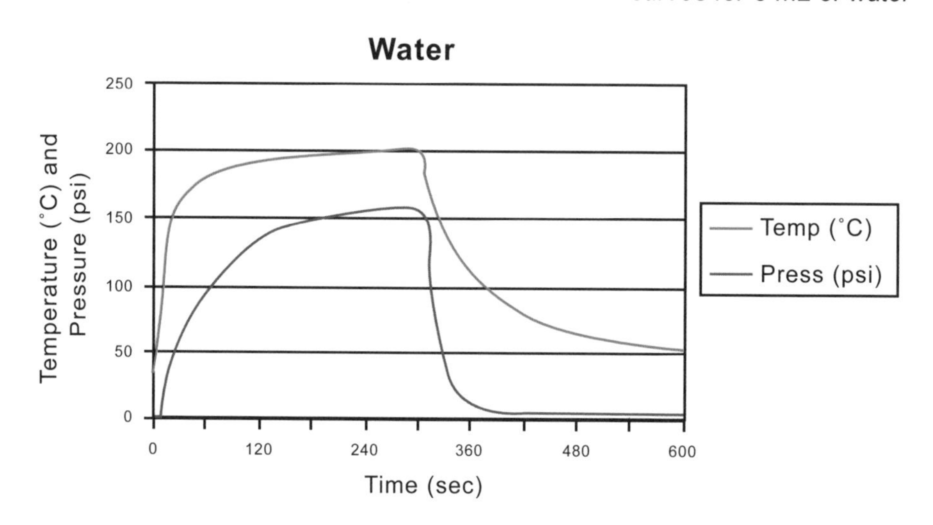

Table 26

Pressures generated at specific temperatures of o-xylene for different volume to head space ratios

Solvent (bp) and Volume	Temperature (°C) /Pressure (psi)	BP + 10°	BP + 25°	BP + 50°
o-Xylene (140)				
1 mL	Temperature	150	165	190
	Pressure	5	10	21
3 mL	Temperature	150	165	190
	Pressure	6	8	20
5 mL	Temperature	150	165	190
	Pressure	6	8	17

Figure 37

Temperature and pressure curves for 3 mL of o-xylene

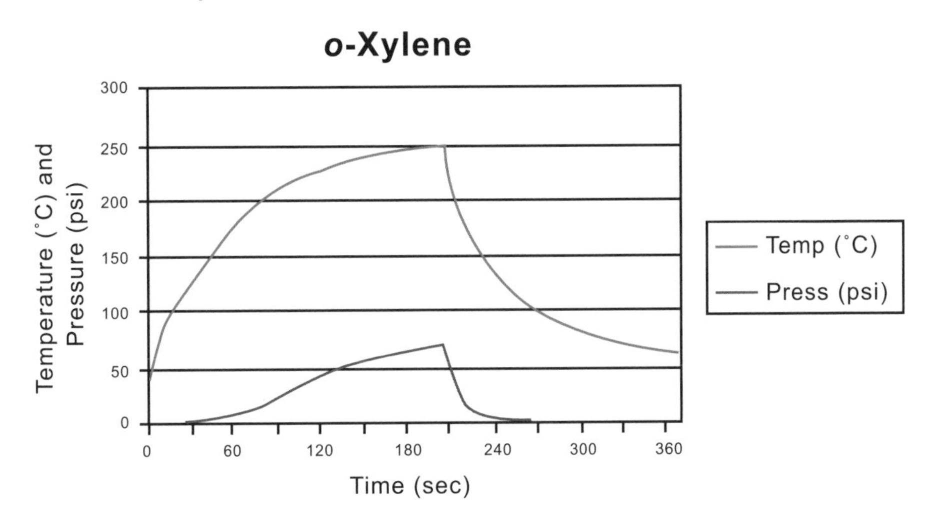

It should be noted that Tables 2-26 and Figures 13-37 only show the temperatures and pressures of neat solvents. There were no reagents present in the pressure vessels. Organic reactions contain many different reagents and catalysts. Their presence can drastically enhance the coupling efficiency of a reaction mixture. The information provided here should only be used as a reference. A typical curve for a chemical reaction will not mirror those shown in any of the above figures. As an example, Scheme 1 shows a Heck reaction that was performed in 0.5 mL DMF with the following programmed method: 60 W, 5 min hold time, 200 °C, 250 psi. Using the same parameters, a control run was performed with 0.5 mL DMF only. Both the reaction and the control run were performed in 10-mL pressure tubes. Figure 38 displays the temperature and pressure curves of each on the same graph for comparison purposes. As the figure shows, the control never reached the set temperature, whereas in the Heck reaction, 200 °C was reached in about 1 minute. Additionally, the pressure achieved in the Heck reaction was almost five times greater than that of the control run. The added reagents greatly increase the polarity of the entire reaction mixture.

Scheme 1

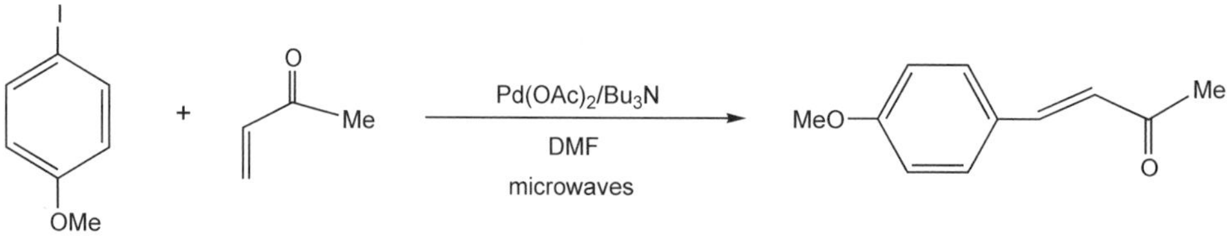

$Pd(OAc)_2/Bu_3N$
DMF
microwaves

Figure 38

Temperature and pressure curve comparison between a Heck reaction and a 0.5 mL DMF control

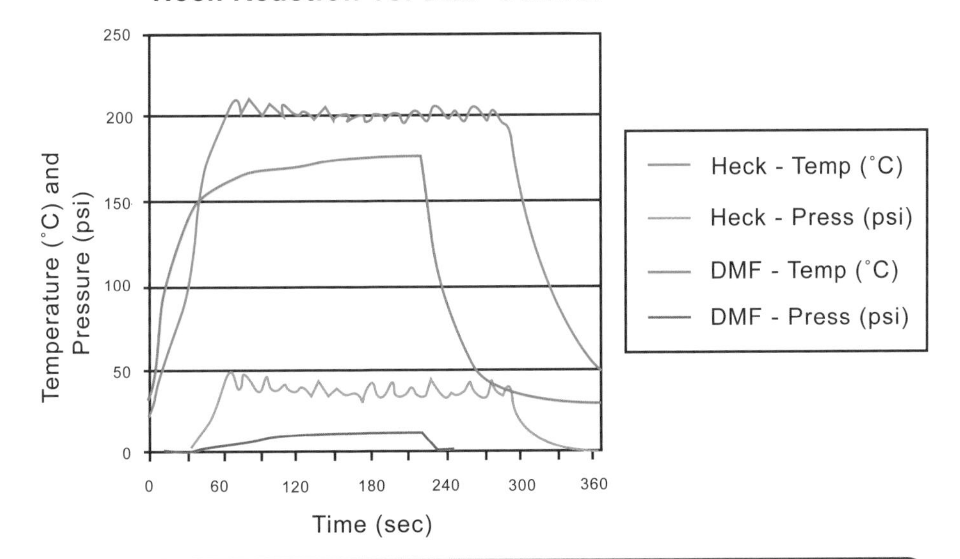

Water becomes a more interesting solvent at higher temperatures and pressures. Under normal conditions, water maintains a very high dielectric constant and persistent hydrogen bonding. As the temperature and pressure of water increases, it begins to act more like an organic solvent. It changes from a very polar liquid to an almost nonpolar one and organic compounds become more soluble. With these enhanced conditions, water has increased acidity, reduced density, and a lower dielectric constant. With microwaves, the supercritical levels of water (T_C = 374 °C, P_C = 218 atm = 3204 psi = 221 bar), where gaseous and liquid water coexist, are not quite reached. Nevertheless, the increased temperatures and pressures can be advantageous for organic synthesis in aqueous media.

Many solvents decompose to hazardous components from prolonged exposure to high temperatures. Prior to choosing an organic solvent, a chemist should be aware of the stability of that solvent at high temperatures. This information is provided in Section 10 (Stability and Reactivity) of the Material Safety Data Sheet (MSDS) for that particular solvent. For example, a few of the more commonly used solvents can degenerate to hazardous components at high temperatures. Dichloromethane, 1,2-dichloroethane, and chloroform are among the chlorine-containing solvents and will decompose to hydrochloric acid (HCl), carbon monoxide (CO), and carbon dioxide (CO_2). In addition, both dichloromethane and chloroform will also yield the highly toxic phosgene (ClCOCl). Dimethylformamide (DMF), dimethylacetamide (DMA), acetonitrile, triethylamine, pyridine, and N-methyl-pyrrolidinone (NMP) all will decompose to carbon monoxide (CO), carbon dioxide (CO_2), and nitrogen oxides (N_xO_y). It should be noted that if DMF becomes discolored, it may cause vessel failures and release toxic fumes. Additionally, pyridine and acetonitrile can produce cyanides. Dimethyl sulfoxide (DMSO) also decomposes to toxic components at high temperatures. It can yield sulfur dioxide (SO_2), formaldehyde (CH_2O), methyl mercaptan (MeSH), dimethyl sulfide (Me_2S), dimethyl disulfide (Me_2S_2), and bis(methylthio)methane ($CH_2(SMe)_2$). Upon exposure to high temperatures, hexamethylphosphoramide (HMPA) will turn a cloudy yellow-orange. Thermal decomposition of HMPA produces toxic fumes of phosphines and phosphorous oxides. These are just a few safety issues that an organic chemist should be aware of when performing high temperature reactions in pressurized vessels.

III. Ionic Liquids

Ionic liquids are becoming promising and useful substitutes for standard organic solvents. Not only are they environmentally benign, but they also possess unique chemical and physical properties.[19] As their name indicates, ionic liquids are only comprised of ions and can also be referred to as fused salts. In general, these fused salts contain one positively charged ion and one negatively charged ion. They have a vast liquid temperature range of almost 300 °C, from –96 °C to 200 °C (unlike water which only has a range of 100 °C). Though they usually consist of poorly coordinating ions, ionic liquids are highly polar, nonvolatile, and readily dissolve both organic and inorganic compounds. These characteristics are all quite beneficial to synthetic organic chemists.

Ionic liquids, also known as fused salts, contain one positively charged ion and one negatively charged ion.

Ionic liquids are either organic salts or mixtures consisting of at least one organic component. They are usually prepared by metathesis of a halide salt of the desired cation with a Group 1 metal or an ammonium salt of the desired anion. Figure 39 shows the most common salts, which are alkylammonium, alkylphosphonium, N-alkylpyridinium, and N,N-dialkylimidazolium cations, respectively. The anion may be organic or inorganic, and there are a number of options to choose from: CH_3COO^-, CF_3COO^-, F^-, Cl^-, Br^-, I^-, BF_4^-, PF_6^-, NO_3^-, $AlCl_4^-$, $FeCl_4^-$, $NiCl_3^-$, $ZnCl_3^-$, and $SnCl_5^-$. Figure 40 exhibits common ionic liquids, some of which are even commercially available.

Figure 39

Common ionic liquid cations

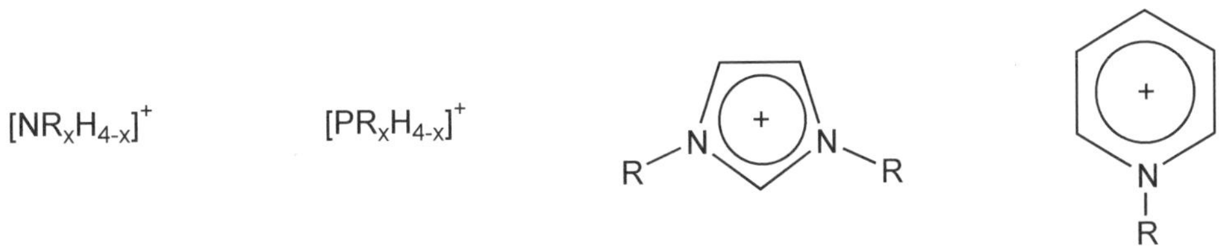

$[NR_xH_{4-x}]^+$ $[PR_xH_{4-x}]^+$

R–N (imidazolium, +) N–R

(pyridinium, +) N–R

Figure 40

Common ionic liquids

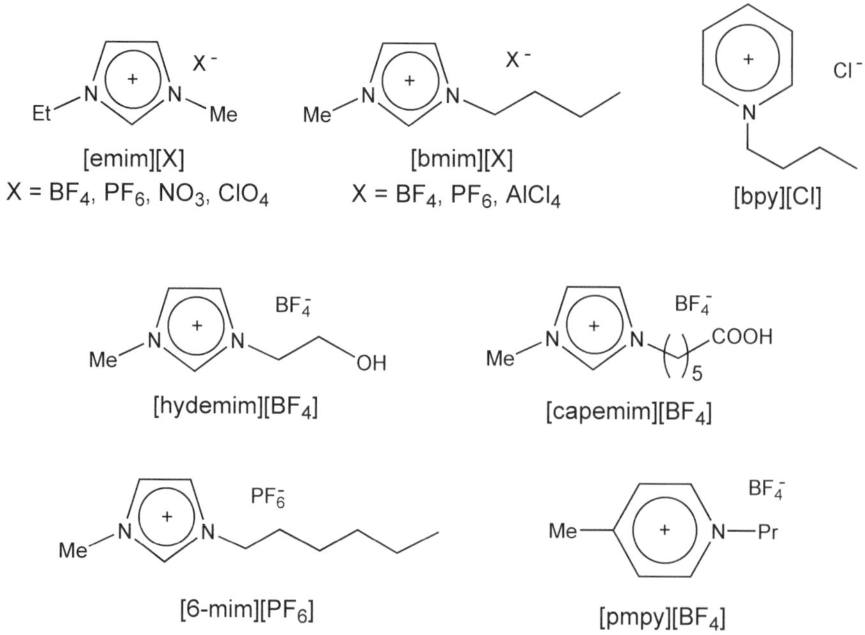

[emim][X]
X = BF_4, PF_6, NO_3, ClO_4

[bmim][X]
X = BF_4, PF_6, $AlCl_4$

[bpy][Cl]

[hydemim][BF_4]

[capemim][BF_4]

[6-mim][PF_6]

[pmpy][BF_4]

Scheme 2

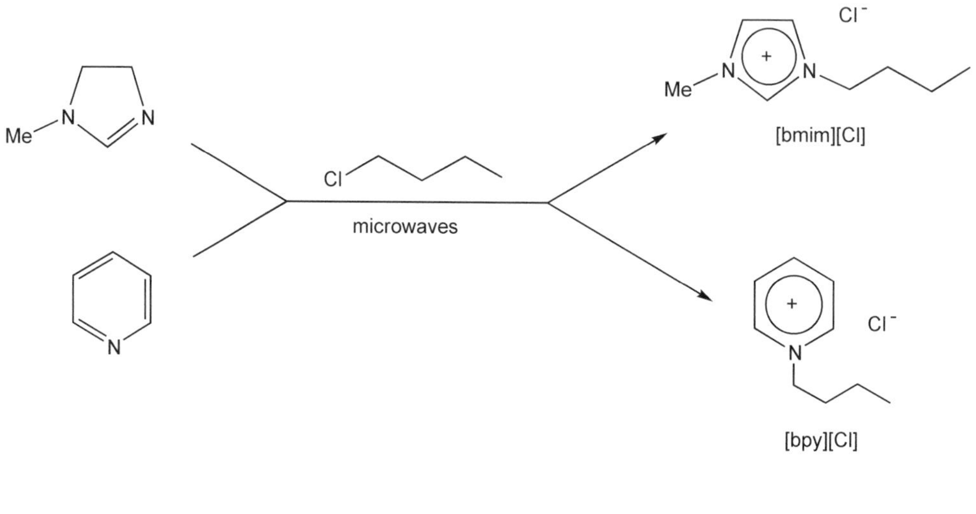

microwaves

[bmim][Cl]

[bpy][Cl]

The conventional preparation of ionic liquids is quite time consuming, as they can require up to 7 days of reflux; thus, microwave irradiation is a preferable method to activate and speed up ionic liquid synthesis.[20,21] Khadilkar et al. synthesized both 1-butyl-3-methylimid-azoliumchloride [bmim][Cl] and 1-butylpyridinium-chloride [bpy][Cl] in 60 minutes and 22 minutes, respectively, with microwave heating (Scheme 2).[21]

Microwave irradiation has also been used to enhance organic reactions in which an ionic liquid is used as the solvent. As discussed in the introduction, the two fundamental mechanisms for transferring energy from microwaves to the substance being heated are either dipole rotation or ionic conduction. Ionic liquids absorb microwave irradiation extremely well and transfer energy rapidly by ionic conduction. Figures 41 and 42 exhibit the temperature and pressure curves for two ionic liquids in 2 mL of hexane (1M [emim][PF_6] and 1M

Figure 41

Temperature and pressure curve comparison between a 1M [emim][PF_6] hexane solution (2 mL) and a 2-mL hexane control

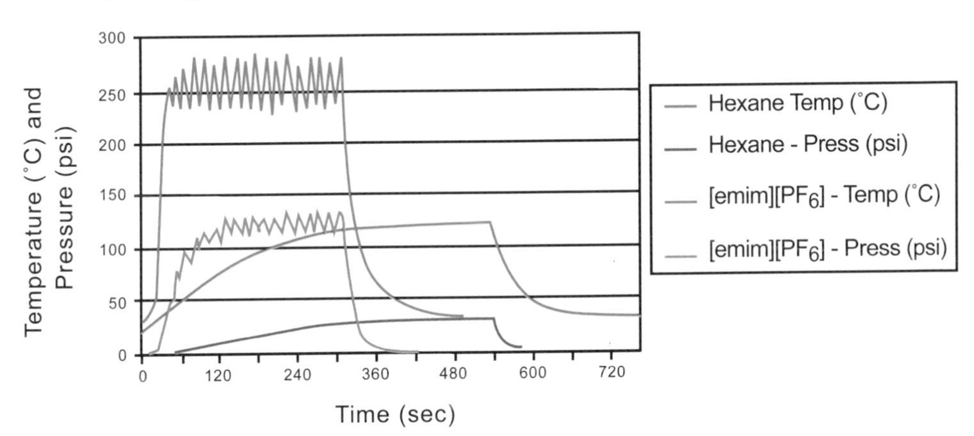

$[emim][BF_4]$, respectively) versus a 2-mL hexane control. Each sample was run with the following programmed method: 100 W, 5 min ramp time, 5 min hold time, 250 °C, 250 psi. All runs were performed in 10-mL pressure tubes. As the graphs indicate, the hexane control did not reach the maximum set temperature, and it needed nine minutes to even reach 120 °C. In both 1M ionic liquid solutions, the temperature was met and exceeded in less than 40 seconds. Additional ionic liquid experimentation in solvents has recently been published by Leadbeater et al.[22]

Figure 42

Temperature and pressure curve comparison between a 1M $[emim][BF_4]$ hexane solution (2 mL) and a 2-mL hexane control

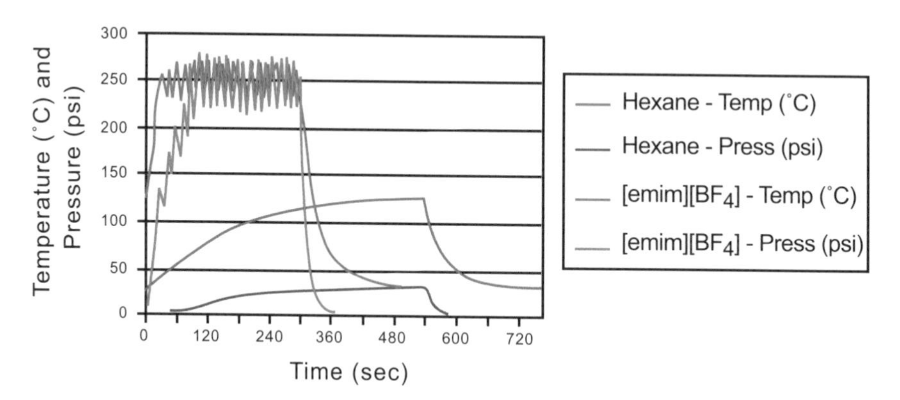

The following 1,3-dipolar cycloaddition is an example of a microwave-enhanced organic reaction in an ionic liquid (Scheme 3).[23] Scheme 4 shows a microwave-induced Knoevenagel condensation reaction between a malonate derivative (EWG = electron withdrawing group) and a grafted ionic liquid phase.[24,25] After the reaction occurs, the ionic liquid phase can be removed and then regenerated for future use.

Scheme 3

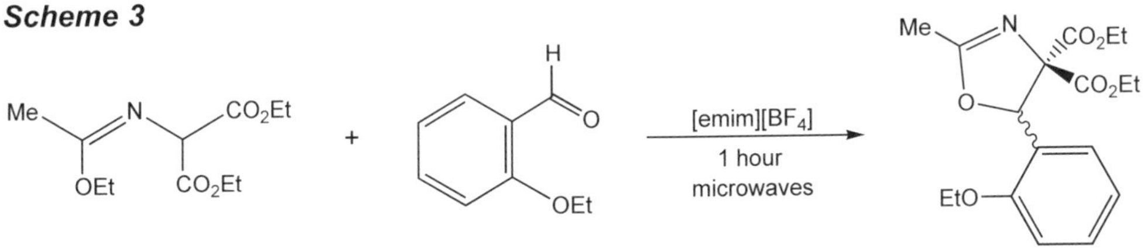

[emim][BF_4]
1 hour
microwaves

Scheme 4

15-60 min
microwaves

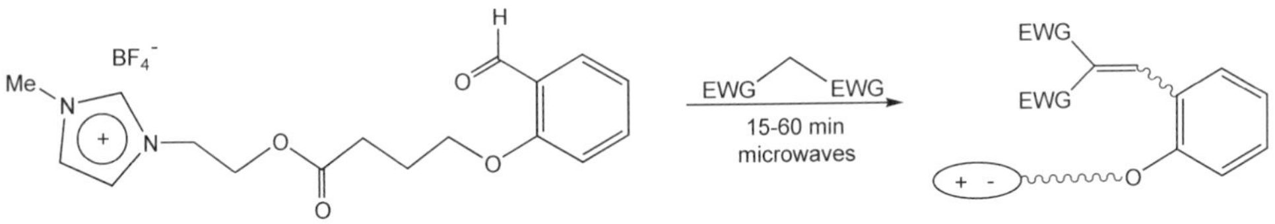

V. Choosing a solvent

An important step before attempting a microwave enhanced organic reaction is choosing a solvent. As discussed earlier in this chapter, the coupling efficiency of a solvent is very important to the outcome of the reaction. The more efficient a solvent is in coupling with the microwave energy, the faster the temperature of the reaction mixture increases. Table 27 contains some common organic solvents that have been categorized as high, medium, or low absorbers. It is a condensed version of Table 1. In addition, a pressurized environment can be very advantageous in performing many different kinds of chemistries. Microwave energy (300 W) will reach and bypass the boiling point (Tables 2-26) of most solvents in a matter of seconds. Using pressurized reaction vessels provide for greater use of the lower boiling point solvents that are normally ignored in conventional high temperature reactions.

The more efficient a solvent is in coupling with the microwave energy, the faster the temperature of the reaction mixture increases.

Table 27

High, Medium, and Low absorbing solvents

Absorbance Level	Solvents
High	DMSO; EtOH; MeOH; Propanols; Nitrobenzene; Formic Acid; Ethylene Glycol
Medium	Water; DMF; NMP; Butanols; Acetonitrile; HMPA; Methyl Ethyl Ketone, Acetone, and other ketones; Nitromethane; *o*-Dichlorobenzene; 1,2-Dichloroethane; 2-Methoxyethanol; Acetic Acid; Trifluoroacetic Acid
Low	Chloroform; Dichloromethane; Carbon Tetrachloride; 1,4-Dioxane, THF, Glyme, and other ethers; Ethyl Acetate; Pyridine; Triethylamine; Toluene; Benzene; Chlorobenzene; Xylenes; Pentane, Hexane, and other hydrocarbons

Choosing a solvent can be a dilemma. The first question to ask is whether high temperature, high pressure, or high energy is needed. When using conventional heat, chemists are usually concerned with the temperature and select a solvent accordingly. Pressure vessels and oil baths are used with higher boiling point solvents to maximize the temperature increase. Achieving high temperatures with microwaves is not the task here;

they can do that with hardly any effort. If a high temperature is required, then choose a solvent that will reach the set temperature. If maintaining a high pressure is all that is desired, then set a high maximum temperature for a low boiling point solvent. The pressure will increase rapidly as the temperature continues to rise above its boiling point. These are classical requirements that microwave energy can easily provide.

High energy is different. It is why microwaves have produced dramatically favorable results and the reason microwave energy is so beneficial to organic synthesis, as opposed to conventional heating which is much slower. The energy transfer in a microwave-assisted reaction is incredibly quick: energy is transferred every nanosecond that it is applied. When performing a microwave reaction, the user can program the power, temperature, time, and, in some cases, a pressure limit. As the temperature reaches the input value, the power is reduced so that the reaction mixture does not bypass the set point. It then stays at a lower level in order to maintain the set temperature throughout the entire reaction. It is the power, or energy, that is the most important variable in a microwave-enhanced reaction. Recent experimentation has shown that simultaneous cooling of the reaction vessel during a reaction will ensure a constant, high power level for direct molecular heating. This has dramatically affected reaction rates and nearly doubled the percent yields of some lower yielding reactions.[18]

Simultaneous cooling of the vessel during a reaction has been shown to nearly double percent yields in some lower yielding reactions.

One would assume that nonpolar solvents (i.e. hexane, benzene, toluene) are generally not used in microwave-assisted organic reactions. In Table 1, these solvents are all at the bottom of the three columns, possessing very low dielectric constants, tan δ values, and dielectric loss values. They do not couple very efficiently to microwave irradiation, and hence, would not heat a reaction very well. Conversely, a nonpolar solvent can act as a heat sink. Reaction mixtures that are temperature sensitive will benefit greatly from this capability. As microwaves are being added to a reaction, the nonpolar solvent, which is not interacting with the irradiation, will help to draw away the thermal heat being produced from the polar reagents. The reaction is still receiving activation energy, but its internal temperature will remain low. Simultaneous cooling of the microwave cavity can benefit this reaction condition and ensure a constant, high power level.

General synthetic organic chemistry rules still apply with microwave-assisted chemical reactions. Regardless of whether one is performing a nucleophilic substitution, electrophilic substitution, or elimination reaction, the type of solvent for each remains the same. There are protic and aprotic solvents, and each of these may or may not be applicable for certain kinds of chemistry. **Protic** solvents have the ability to solvate or interact with both cations and anions, whereas **aprotics** can only solvate cations. The solvents of each type are interspersed throughout Table 1. Chemists must use a combination of the previously discussed information from this chapter in order to determine their reaction conditions.

Protic solvents solvate or interact with both cations and anions. Aprotic solvents only interact with cations.

Nucleophilic substitution reactions (S_N2, S_N1, etc.) depend considerably on solvent effects for their success. Whether a transition state is stable or unstable in a solvent greatly influences the outcome of the reaction. Additionally, for S_N2 reactions, stabilization or destabilization of the reactant nucleophile is also a major factor. As illustrated in Scheme 5, S_N2 reactions can be categorized into four types (I, II, III, and IV), and they generally require aprotics (hexane, benzene, Et_2O, $CHCl_3$, ethyl acetate, acetone, HMPA, DMF, DMSO, acetonitrile).[26] Protic solvents are disfavored in these reactions, since the ground state energy level of the attacking nucleophile is lowered by solvation. In other words, the nucleophile is stabilized by solvation and, therefore, less reactive towards the electrophile. The polarity of the aprotic solvent is also important. The only type in Scheme 5 that succeeds in highly polar solvents is Type II because the reactants are uncharged. The reactions of Types I, III, and IV have at least one charged reactant, and the reaction is actually hindered by a very polar solvent.

Scheme 5

Type I $$R{-}I + {}^{-}OH \longrightarrow R{-}OH + I^{-}$$

Type II $$R{-}I + NMe_3 \longrightarrow R{-}\overset{+}{N}Me_3 + I^{-}$$

Type III $$R{-}\overset{+}{N}Me_3 + {}^{-}OH \longrightarrow R{-}OH + NMe_3$$

Type IV $$R{-}\overset{+}{N}Me_3 + SH_2 \longrightarrow R{-}\overset{+}{S}H_2 + NMe_3$$

The stability of the transition state in S_N1 reactions is extremely important. The solvent used strongly influences this stability. S_N1 reactions generally occur more rapidly in highly polar protic solvents than in aprotic and nonpolar solvents, though there are exceptions. The energy level of the transition state that leads to the carbocation intermediate is lowered by solvation. Imagine solvent molecules orienting themselves around the cation in such a manner that the electron-rich ends of the solvent dipoles face the positive charge. This is described in the Hammond postulate, which states that any factor stabilizing the intermediate carbocation should increase the rate of reaction.[26] Alcohols, water, and formic acid are good solvents for S_N1, but HMPA, DMF, and DMSO have also been known to work well since they are polar.

The solvent effects necessary for successful electrophilic substitution reactions are slightly different from those just discussed for nucleophilic substitutions. Mechanistically, S_E2 is analogous to S_N2 where a new bond forms as the old one breaks. One difference is in the solvent preference. The reaction rate increases as solvent polarity increases. The mechanism of the S_E1 reaction is analogous to the S_N1 reaction: slow ionization occurs as the bond breaks, followed by new bond formation. As with the S_N1 reaction, S_E1 reactions are faster and more successful in highly polar solvents.

Elimination reactions (a double bond is formed by either simultaneous (E2) or sequential (E1) leaving group departure and proton abstraction) are analogous to nucleophilic substitutions, E2 to S_N2 and E1 to S_N1. Interestingly, in many cases, an E2 reaction will compete with an S_N2 in the same reaction mixture. Both contain a nucleophile that will happily abstract a proton or attack an electrophilic carbon. In general, increasing the polarity of the solvent will favor substitution over elimination. Alternatively, elimination will be more favorable if the solvent is nonionizing and in the presence of a strong base. For E1 reactions, a more polar solvent will enhance the rate of the mechanism, especially one that involves an ionic intermediate, as has been the case with the other two-step reaction mechanisms (S_N1, S_E1).

Thus, solvents play an extremely significant role in microwave-enhanced organic chemistry. Using the dielectric loss values from Table 1, coupled with the general rules of organic chemistry previously described in this chapter, chemists can now develop the specific conditions that will optimize their synthetic endeavors.

Chapter 3
Chemical Reactions in the Presence and Absence of Solvent

Microwave irradiation is not only applicable to standard solution-phase work, but to solid-phase and solvent-free systems as well. Many synthetic methods can be executed by at least one of these systems, though one may have to experiment in order to find the optimal conditions. This chapter discusses the different types of reaction conditions that can be used successfully with microwave irradiation and should be utilized in conjunction with Chapter 4 for synthetic applications. *Note: The reader should assume that all reaction schemes shown in this chapter utilize microwave irradiation. In multi-step schemes, the use of microwaves is indicated by the word "microwaves" on the arrow.*

The two main types of conditions used for chemical reactions, those run in the presence of solvent and those run in a solventless environment, are equally important and both can benefit from microwave heating.

I. Reactions in the presence of solvent

Solution-phase reactions performed in the presence of solvent can be either homogeneous or heterogeneous. Homogeneous reactions include your standard organic reactions in which all reagents are dissolved in the solvent. Microwave irradiation has been used extensively and successfully with homogeneous solution-phase reactions. Chapter 4 provides an in-depth review of the many homogeneous synthetic applications that have been enhanced with microwaves.

Heterogeneous reactions in solution involve insoluble solids that are used as reagents, catalysts, or supports. These include transition-metal and Lewis acid catalysts, non-dissolvable salts, and solid-phase resins (beads, lanterns, crowns, pins). These types of transformations are widely used and highly successful. One of the main disadvantages in traditional heterogeneous reactions are the long reaction times that are required for their completion, which is largely due to the insolubility. Use of microwave irradiation has been shown to drastically speed up these reactions, allowing for productive high throughput synthesis.

The following chapter on synthetic applications extensively covers heterogeneous reactions involving transition-metal catalysts, as well as those that utilize Lewis acids and other insoluble salts. It also provides a few examples of solid-phase reactions.[188,204,609,631] The remainder of this section will provide a more detailed review on microwave-enhanced reactions that have been performed on a solid-phase resin.[27-42]

Combinatorial chemistry on solid-phase supports was first applied to peptide synthesis. It made sense to use peptides in the first microwave-assisted, solid-phase reaction. Traditional peptide hydrolysis requires high temperature 6M HCl for at least 24 hours. In 1988, Yu et al. performed a successful hydrolysis in seven minutes in a domestic microwave oven (Scheme 6).[27]

The same group, four years later, performed peptide coupling reactions in quantitative yields with microwave irradiation (Scheme 7).[28]

Scheme 6

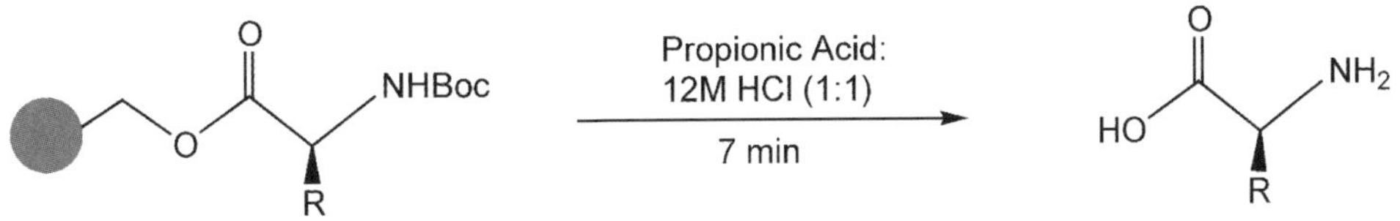
O
NHBoc
O
R
Propionic Acid:
12M HCl (1:1)
7 min
O
NH2
HO
R

Scheme 7

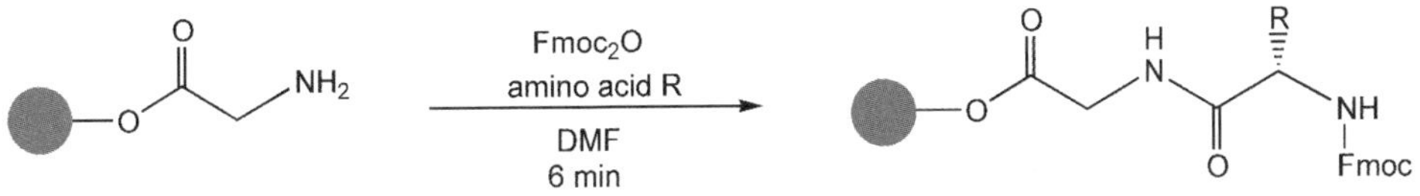
O
NH2
O
Fmoc2O
amino acid R
DMF
6 min
O
H
N
R
O
NH
Fmoc
O

Sigmatropic rearrangements are important pericyclic reactions that involve the formation of new carbon–carbon bonds. Sometimes, conventional methods require very long reaction times. Microwave irradiation has been used to facilitate these transformations, and this is outlined in the next chapter. Claisen rearrangements have been successfully performed on a solid-phase resin coupled with microwave heating. Scheme 8 illustrates the rearrangement of resin bound O-allylic

Scheme 8

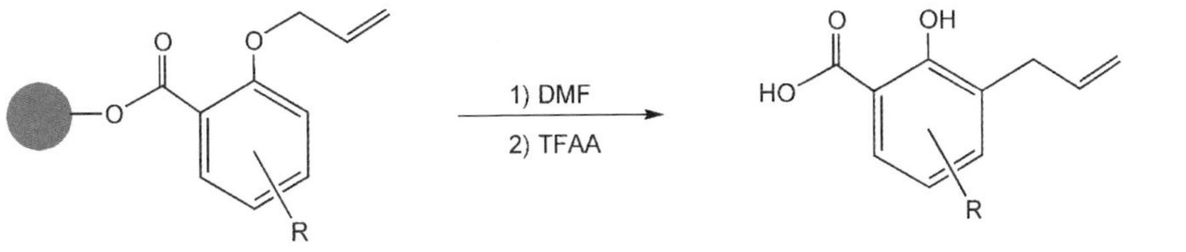
O
O
O
R
1) DMF
2) TFAA
O
OH
HO
R

Microwave: 4-6 min, 68-92% yield
Conventional: 10-16 hours, 60-90% yield

aryl ethers to *ortho*-allylic salicylic acid derivatives, where the allyl, hydroxyl, and carboxylic acid groups are adjacent to each other.[38] These compounds are difficult to synthesize with traditional aromatic substitution reactions.

Strohmeier and Kappe have performed an important example of microwave-assisted, solid-phase parallel synthesis.[29] Enones, as well as their 1,3-dicarbonyl intermediates, are important building blocks for heterocyclic scaffolds that are used in pharmaceutical drug design. Conventional solid-phase methods require multiple steps and long high-temperature reaction times. In this microwave-enhanced, two-step procedure, acetoacetylation of a polystyrene Wang (PS-Wang) resin to a functionalized β-ketoester, followed by Knoevenagel condensation with an aldehyde, yields enones in less than one hour (Scheme 9).

Scheme 9

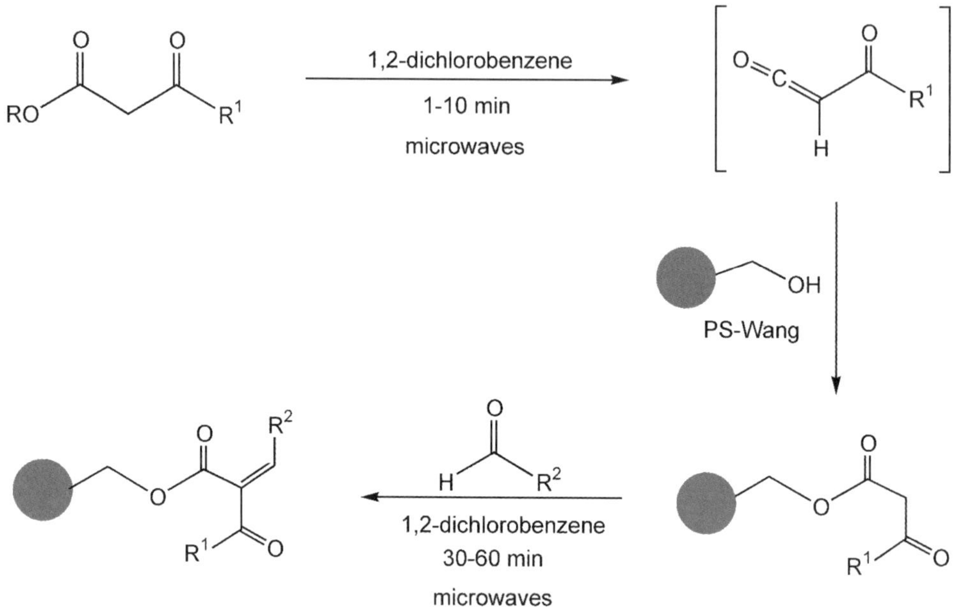

1,2-dichlorobenzene
1-10 min
microwaves

PS-Wang

1,2-dichlorobenzene
30-60 min
microwaves

Glass and Combs have also performed rapid parallel synthesis on solid-phase resins with microwave irradiation.[31] They have examined the utility of a "safety catch" sulphonamide linker to produce large libraries of diverse amides and ureas. These safety catch linkers are highly stable through a given synthetic sequence until cleavage from the resin is necessary. The linker must first be activated before nucleophilic displacement of the substrate can occur. Conventional displacement of the substrate requires a very strong nucleophile and limits library diversity. The use of microwave irradiation allows for any nucleophile to displace the resin, including weak ones. Scheme 10 shows a facile biaryl urea synthesis, which includes a Suzuki coupling reaction, linker activation via alkylation, and subsequent cleavage with diisopropyl amine (DIPA).

Scheme 10

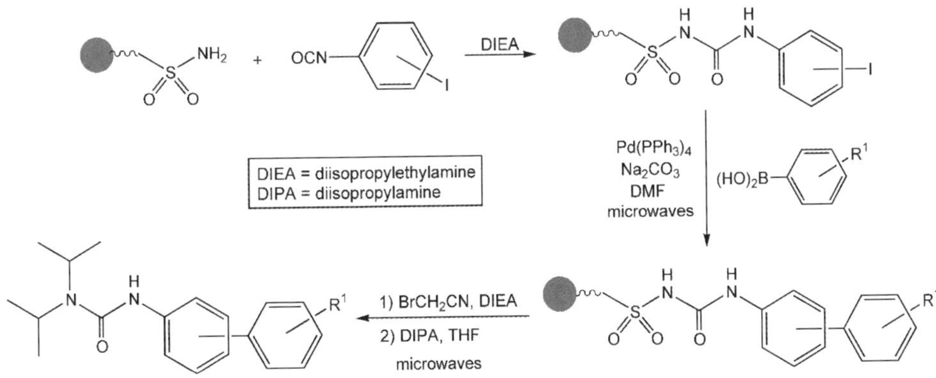

Microwave-assisted, solid-phase syntheses are not restricted to spherical bead polymer resins. Scharn et al. used a planar cellulose membrane to synthesize a parallel library of 8000 1,3,5-triazines via nucleophilic substitution reactions.[40] The planar membrane is composed of an array of "spots" that are individually derivatized. Scheme 11 illustrates how an amino-functionalized spot is doped with cyanuric chloride and then diversified with different amines by microwave heating. The second nucleophilic substitution requires five hours of thermal heat for completion, but with microwave irradiation, an entire library can be synthesized in six minutes.

Scheme 11

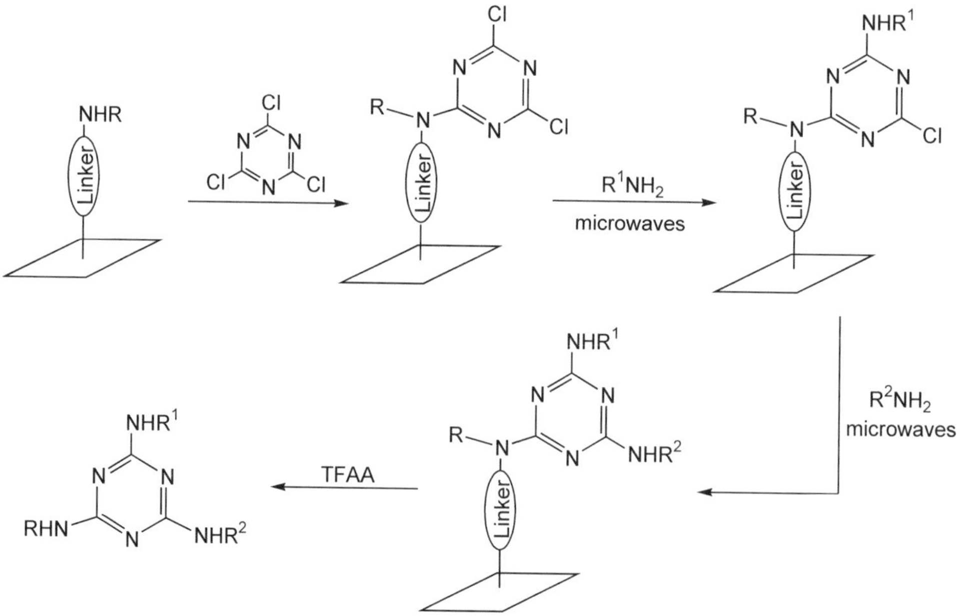

II. Solvent-free reactions

Reactions performed in a solvent-free environment are becoming more prevalent in organic chemistry. An increasing need for less hazardous reaction conditions and environmentally safe procedures, or green chemistry, has led chemical synthesis in this direction. Microwave irradiation has been used extensively in solvent-free reactions.[5,8,16,20,43-181] There are three main types of solvent free reactions: reaction mixtures adsorbed onto mineral oxides, phase transfer catalysis (PTC), and neat reactions.

This section will identify and provide examples for each type. For an overview of a wider range of different chemical transformations that can be performed solventless, the reader should consult Chapter 4, as there are over 100 additional solvent-free references.

An increasingly popular solvent-free method is to adsorb reagents onto mineral oxides. The reagent is first dissolved in an appropriate volatile solvent. After the mineral oxide (alumina, silica gel, clay, or zeolites) is added, the solvent is removed by evaporation. The impregnated solid support is then irradiated with microwaves in "dry media". Upon completion of the reaction, a solvent is added to extract the product(s) from the support. Choice of solid support depends on the type of reaction a chemist is going to perform. Alumina can act as a base, but if a stronger one is needed, potassium fluoride on alumina is extremely basic. Silica gel naturally acts as a weak acid, while some of the montmorillonite clays provide acidities near sulfuric and nitric acids. As a whole, this solid-state application will greatly reduce the amount of solvent used that eventually needs to be properly disposed of and will minimize potentially hazardous reaction conditions.

Kidwai and co-workers have done extensive research in solvent-free reaction chemistry.[37,104-113,276,290] Scheme 12 shows an example of a microwave-enhanced synthesis to N-acylated cephalosporin derivatives.[37] Cephalosporanic acid and a heterocyclic carboxylic

Scheme 12

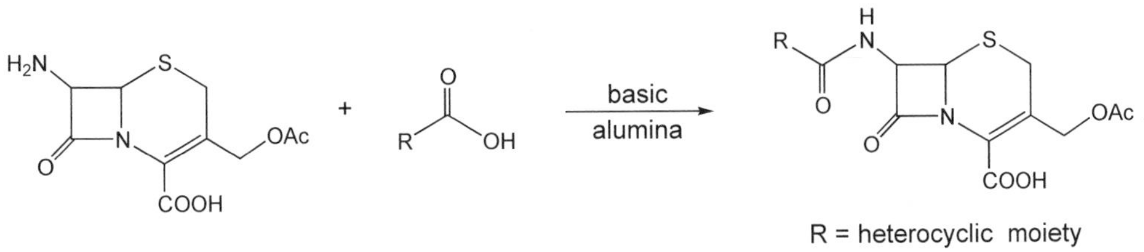

basic alumina

R = heterocyclic moiety

Scheme 13

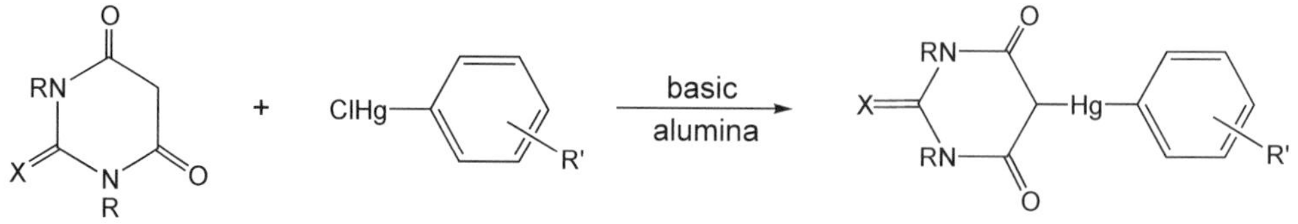

basic alumina

X = O,S
R = H,Ph
R' = H,Cl,OMe

acid were adsorbed onto basic alumina and irradiated with microwaves for 2 minutes to yield the antibacterials in 82-93% yield. With thermal heat, this reaction can take anywhere from two to six hours and provides much lower yields. Another reaction performed on basic alumina is shown in Scheme 13.[105,113] Barbituric and thiobarbituric acid derivatives are adsorbed onto the alumina with substituted arylmercuric chlorides to yield biologically active fungicides.

Kabalka and co-workers have also explored solventless, microwave-enhanced reactions on dry media.[99-101,628-629] Sonogashira coupling reactions are a palladium-catalyzed reaction between terminal alkynes and an aryl halide. These reactions typically employ a solvent and an amine, which produce environmental burdens. Scheme 14 illustrates a Sonogashira coupling that was performed on potassium fluoride/alumina doped with a palladium/copper iodide/triphenylphosphine mixture. The arylalkynes were synthesized in very high yields (82-97%).[99] Another type of coupling reaction that can be performed in a solvent-free environment is Glaser coupling. This copper-catalyzed coupling of two terminal alkynes produces diacetylene derivatives, which are very important in the polymer and material science industries. Phenylacetylene and copper chloride on potassium fluoride/alumina, coupled with microwave irradiation, give diphenylbutadiyne in a 75% product yield (Scheme 15).

Scheme 14

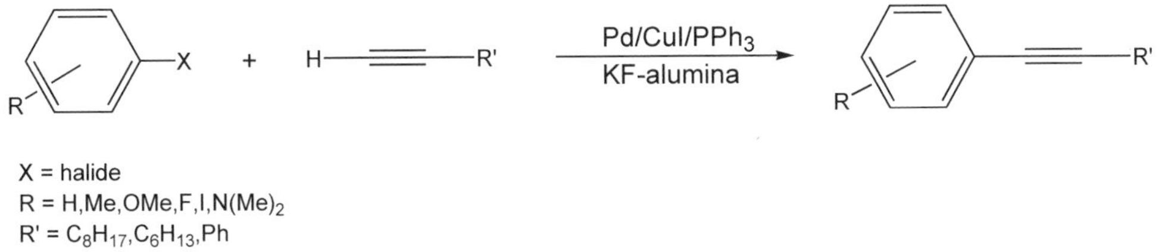
X
+
H
R'
Pd/CuI/PPh3
KF-alumina
R
R'
X = halide
R = H,Me,OMe,F,I,N(Me)2
R' = C8H17,C6H13,Ph

Scheme 15

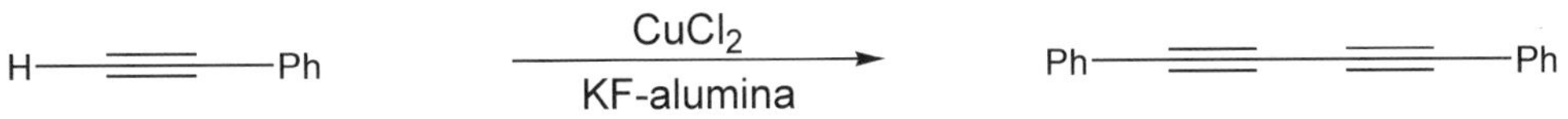
H
Ph
CuCl2
KF-alumina
Ph
Ph

The Baylis-Hillman reaction is an important carbon–carbon bond forming reaction that forms multifunctional molecules. In this reaction, an aldehyde reacts with an electron deficient alkene to yield allylic alcohol derivatives. Isomerization of acetylated Baylis Hillman adducts will yield (E)-trisubstituted alkenes, which are often difficult to synthesize. Microwave irradiation of the functionalized acetates on montmorillonite K10 clay yields trisubstituted alkenes in 13 minutes (Scheme 16).[130] The clay acts as a catalyst, since only starting material is recovered in its absence.

Scheme 16

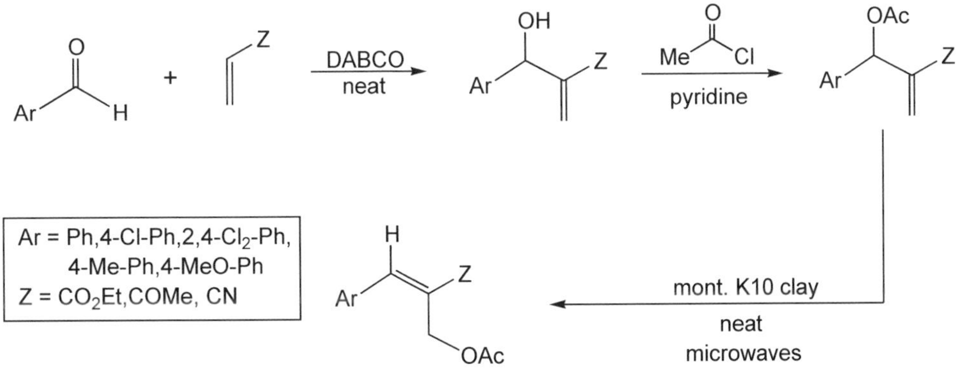
O
Ar
H
+
Z
DABCO
neat
OH
Ar
Z
O
Me
Cl
pyridine
OAc
Ar
Z
Ar = Ph,4-Cl-Ph,2,4-Cl2-Ph,
4-Me-Ph,4-MeO-Ph
Z = CO2Et,COMe, CN
H
Ar
Z
OAc
mont. K10 clay
neat
microwaves

Another pioneer in microwave-assisted solvent-free reactions is Andre Loupy.[5,8,45-63,297,310,351,416,425,439,472,491,501,505-507,522,568,590,607,657,708,709] One important reaction that is used frequently in natural product syntheses is the Beckmann rearrangement. This reaction rearranges ketoximes to amides or lactams in the presence of acid. Traditionally, very strong acids are used to promote the rearrangement. Loupy and co-workers have performed facile Beckmann rearrangements on montmorillonite K10 clay under microwave irradiation in high yields (68-96%) (Scheme 17).[60] Another microwave reaction performed by Loupy et al. in a solventless environment is carbohydrate glycosylation. Scheme 18 illustrates the glycosylation of peracetylated D-glucopyranose with decanol.[54]

Scheme 17

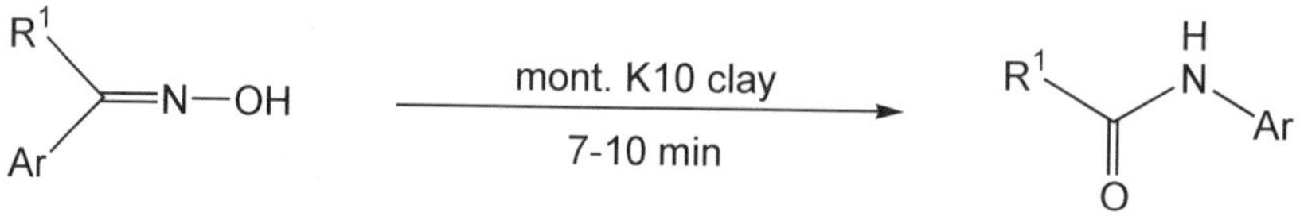

R^1(Ar)C=N—OH → (mont. K10 clay, 7-10 min) → R^1C(=O)NH—Ar

R^1 = Me,Ph

Ar = Ph,p-OMeC_6H_4,p-ClC_6H_4, p-$NO_2C_6H_4$

Microwave: 68-96% yield
Conventional: 17-93% yield

Scheme 18

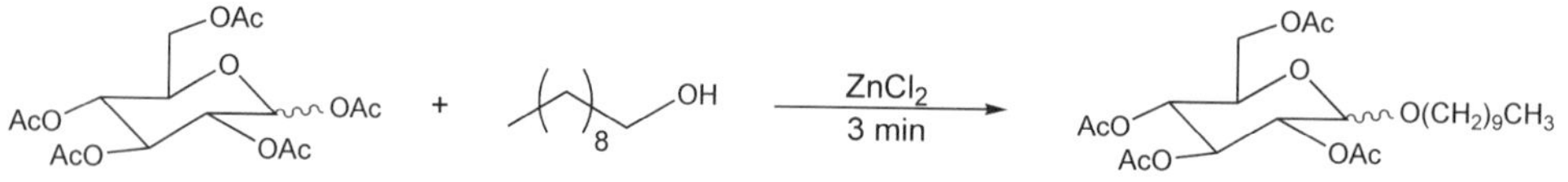

Peracetylated D-glucopyranose (AcO, AcO, OAc, OAc, OAc) + decanol (OH, 8) → ($ZnCl_2$, 3 min) → glycoside with $O(CH_2)_9CH_3$

Microwave: 3 min, 74% yield
Conventional: 5 hours, 25% yield

Varma and co-workers have performed extensive research on microwave-assisted, solvent-free reactions in numerous areas including oxidations, reductions, protections, deprotections, and condensations.[20,84-98] Many of these are discussed in Chapter 4 and include additional references. Another area of interest is the enamine-mediated approach to isoflav-3-ene synthesis. Enamines are traditionally synthesized via azeotropic removal of water and usually require an initial acid catalyst. Scheme 19 shows a microwave-enhanced, solvent-free, one-pot synthesis to isoflav-3-ene derivatives, which takes place in only seven minutes.[98] An efficient microwave-induced tetrahydroquinolone synthesis effected on clay is completed in only two minutes (Scheme 20).[96]

Scheme 19

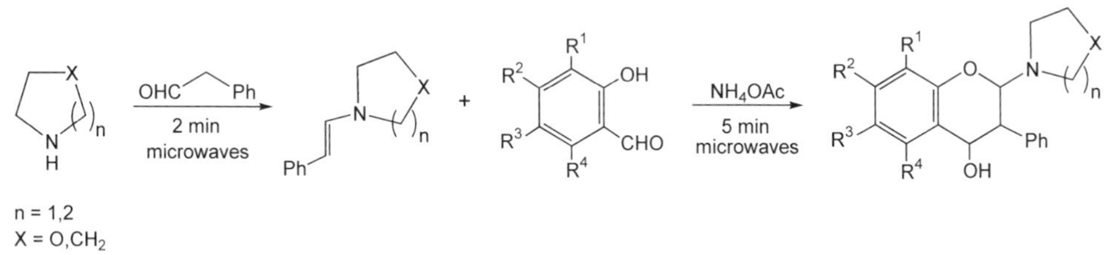

X N H n
OHC Ph
2 min
microwaves
N X n Ph
+
R^1 R^2 R^3 R^4 OH CHO
NH$_4$OAc
5 min
microwaves
R^1 R^2 R^3 R^4 O N X n Ph OH

n = 1,2
X = O,CH$_2$

Scheme 20

NH$_2$ O R
mont. K10 clay
2 min
H N O R

R = H,Me,OMe,Cl,Br,NO$_2$

Didier Villemin is yet another researcher who has examined reactions in dry media extensively. Metallophthalocyanines have become important molecules in the material science industries, as they are stable to strong acids and bases, as well as high temperatures. Traditional synthetic routes to phthalocyanines require long reaction times and very high temperatures. Villemin and co-workers have performed one-step metallophthalocyanine syntheses on clay, zirconium phosphate, and encapsulated in zeolite via microwave irradiation (Scheme 21).[132] These reactions were completed in only five minutes and in quantitative yields.

Scheme 21

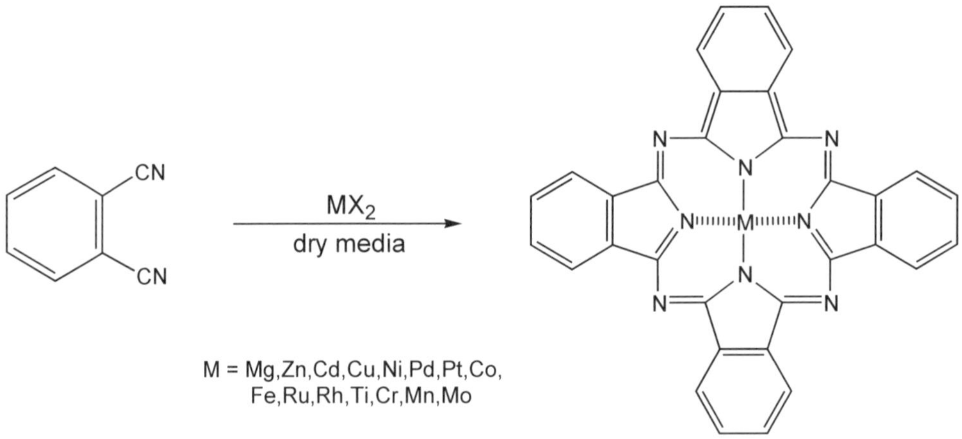

CN
CN
MX_2
dry media
M = Mg,Zn,Cd,Cu,Ni,Pd,Pt,Co,
Fe,Ru,Rh,Ti,Cr,Mn,Mo

Solid-liquid-phase transfer catalysis is another type of solvent-free reaction. With this method, a reagent acts as both a reactant and an organic phase. Microwave irradiation has been used extensively in these types of reactions.[8,50,66,351,472,483,490,491,505,593] An inexpensive and useful phase transfer catalyst (PTC) is polyethylene glycol (PEG). Medium to high molecular weight PEG is a solid at room temperature, but at 50 °C, it melts to become a liquid. At temperatures above 50 °C, derivatized PEG can be used as a soluble polymeric support in the solution phase, but when cooled to room temperature, it becomes solid and provides for simple purification. Scheme 22 exhibits a PEG-supported alkylation of a Schiff base to aminoacid derivatives under microwave irradiation in 75-98% yield.[166]

Scheme 22

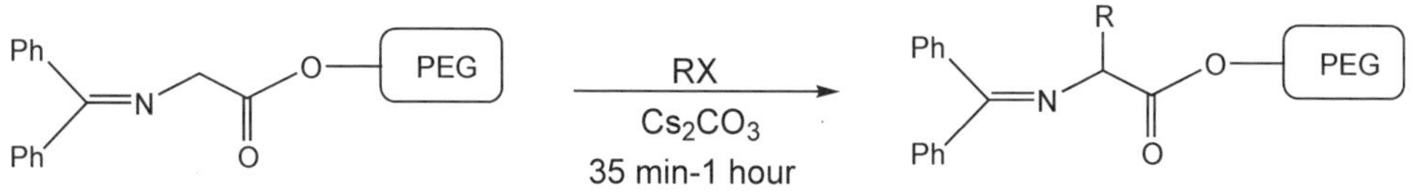

RX, Cs_2CO_3, 35 min-1 hour

Another useful PTC for microwave-assisted reactions is poly(styrene-*co*-allyl alcohol) (Ps-OH). This support possesses the properties of both PEG and polystyrene. Vanden Eynde and Rutot rapidly synthesized heterocyclic compounds via supported β-keto esters, with the first step only taking five minutes (Scheme 23).[41] The parent polymer can be regenerated from the resulting acylated polymer by saponification.

Scheme 23

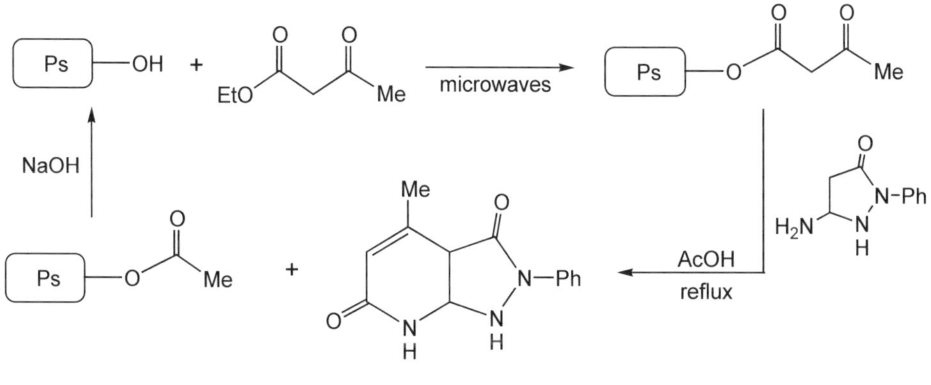

microwaves; AcOH, reflux; NaOH

Andre Loupy has also done some interesting research involving phase transfer catalysis coupled with microwave irradiation.[8,50,62,351,472,491,505,593] β-Elimination of halogenated precursors, with potassium *t*-butoxide/tetrabutylammonium bromide (KO*t*Bu/TBAB) as the PTC, provides a new route to ketene O,O- and S,S-acetals (Scheme 24).[62] Compared to both conventional and ultrasonic methods, microwave irradiation produced much larger product yields.

Scheme 24

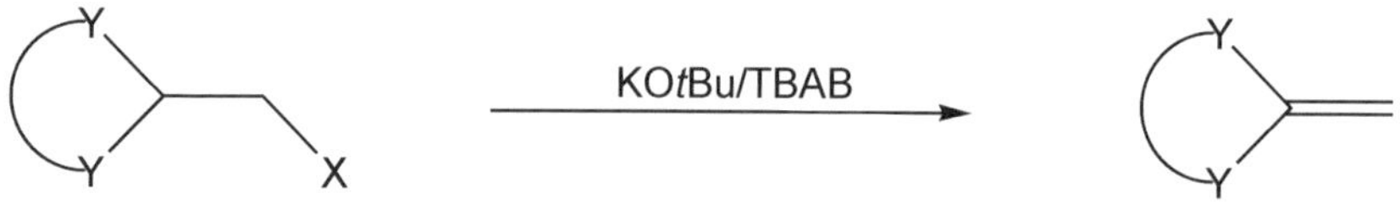

KO*t*Bu/TBAB

X = Cl,Br,I
Y = O,S

Microwave: 25 min, 79% yield
Ultrasound: 25 min, 45% yield
Conventional: 25 min, 15% yield

Another area of interest includes the synthesis of furan diethers. These types of compounds constitute a large percentage of the derivatives that make up biomass, a renewable source of natural products. Loupy and coworkers developed two methods of microwave-assisted phase transfer catalysis for furan synthesis, solid-liquid PTC (solid KOH and Aliquat 336) and liquid-liquid PTC (aqueous KOH and Aliquat 336).[351] Scheme 25 shows the reaction between 2,5-furandimethanol and an alkyl halide by both PTC methods. Phase transfer catalysis can also benefit the reaction between furfuryl alcohol and a dihalide (Scheme 26).

Scheme 25

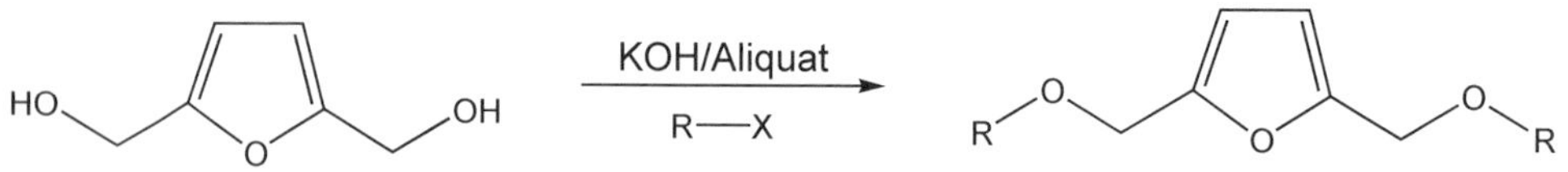

KOH/Aliquat
R—X

Microwave: 10 min, 63-94% yield
Conventional: 30 min, 41-89% yield

Scheme 26

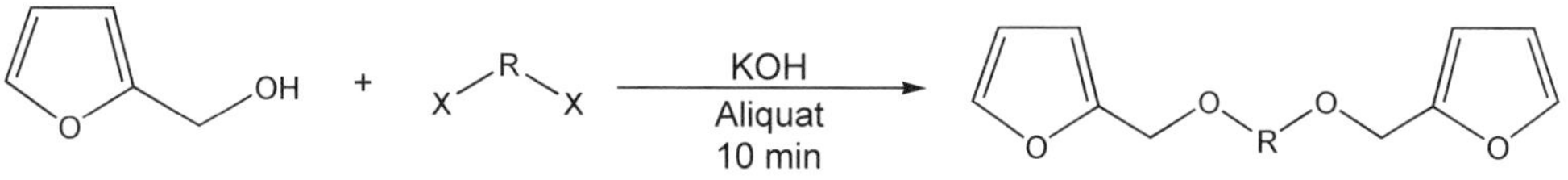

KOH
Aliquat
10 min

Performing a reaction neat under microwave irradiation is the third type of solvent-free reaction. With this method, neither a mineral oxide nor a PTC is used, and the liquid or solid reagents are used directly from their containers with no dilutions. One interesting neat reaction utilizes Lawesson's reagent, which transforms a carbonyl moiety into its thio analog. Scheme 27 exhibits the microwave-induced conversion of amides to thioamides in six minutes.[88,165] An additional microwave example converts coumarins and other lactones to their thio derivatives in only 3 minutes with quantitative product yields (Scheme 28).[88]

Scheme 27

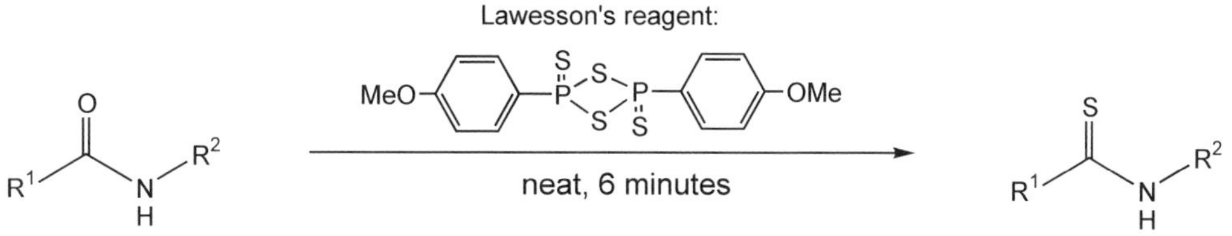

Lawesson's reagent:
neat, 6 minutes

Scheme 28

Lawesson's reagent
neat, 3 minutes

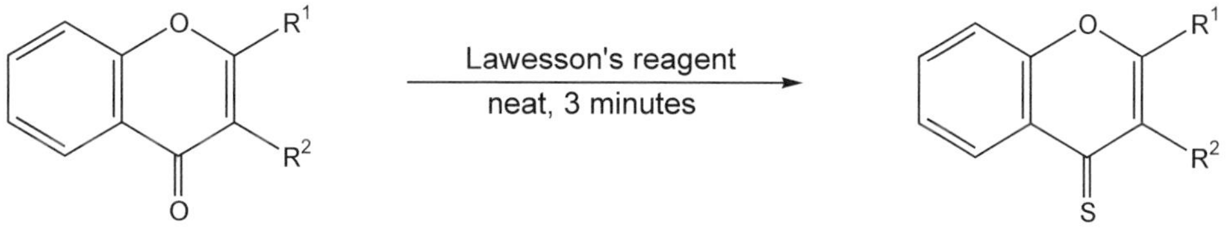

Substituted 2-oxazolines are important heterocyclic intermediates used in drug discovery. Classical syntheses of these compounds require high temperatures, azeotropic water removal, and multi-step procedures. With microwave irradiation, 2-oxazolines are synthesized from the cyclodehydration reaction between a carboxylic acid and α,α,α-tris(hydroxymethyl) methyl amine without any solvent or solid support in 2-5 minutes (Scheme 29).[57]

Scheme 29

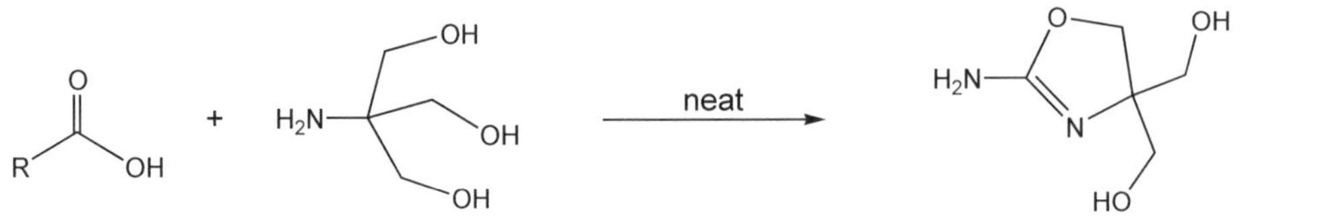

neat

Microwave: 2-5 min, 80-95% yield

Both Diels-Alder and 1,3-dipolar cycloadditions benefit from microwave-assisted neat conditions, as they require long reaction times and very high thermal temperatures. In a solventless environment, vinylpyrazoles react with substituted alkynes to yield non-aromatic cycloadducts via microwave irradiation in 15 minutes (Scheme 30).[214] Schemes 31 and 32 illustrate successful 1,3-dipolar cycloadditions that yielded heterocycles in very high product yields.[8,364]

Scheme 30

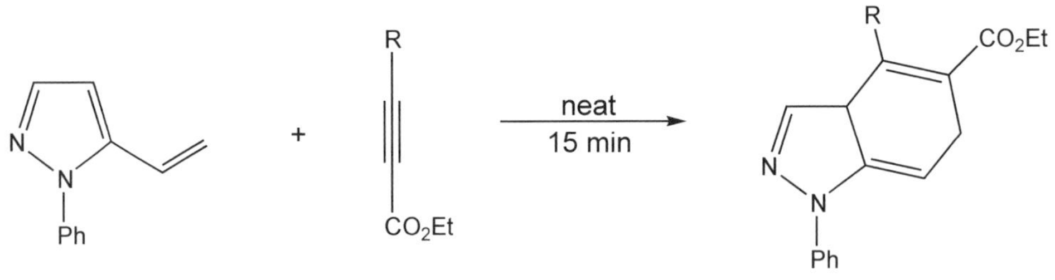

neat
15 min

Scheme 31

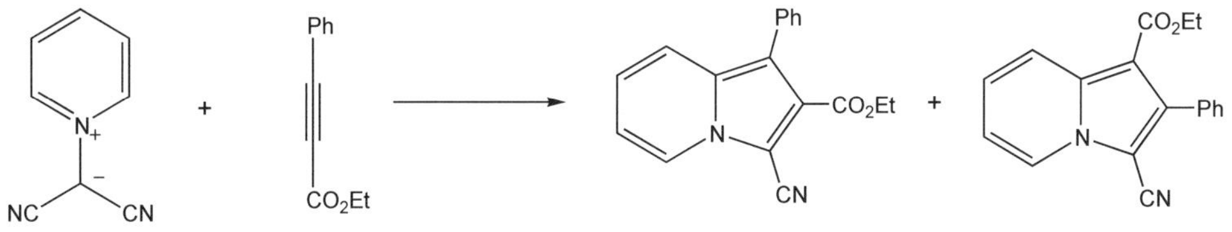

Microwave: neat, 25 min, 87% yield
Conventional: neat, 25 min, 61% yield
Conventional: DMF, 25 min, 15% yield

Scheme 32

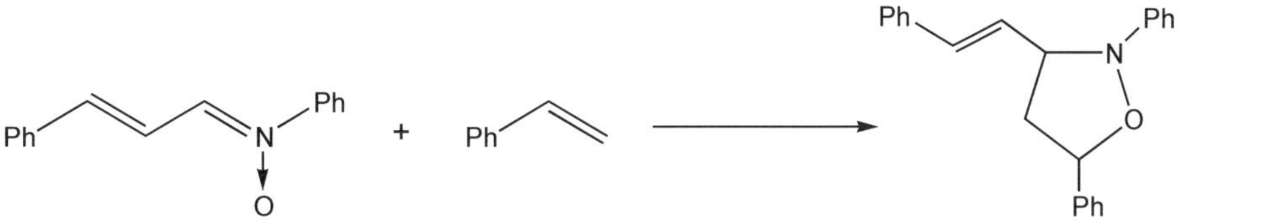

Microwave: 6 min, 90% yield
Conventional: 34 hours, 80% yield
Ultrasound: 1 hour, 87% yield

Thus, the two main types of conditions used for chemical reactions, those run in the presence of solvent and those run in a solventless environment, are equally important and both can benefit from microwave heating. We have seen that microwave irradiation is not only applicable to standard homogeneous reaction mediums, but to solid-phase systems as well. Most synthetic methods can be executed by at least one of these systems. In conjunction with the following synthesis chapter, a chemist can now develop optimal and efficient synthetic routes.

Chapter 4
Synthetic Applications

In the past five years, there has been an increased demand for large collections of novel drug targets. The long reaction times that are required for conventional heating have led to the advent of new technologies, including combinatorial and parallel chemistry. Combinatorial chemistry allows a chemist to synthesize large libraries of molecules by varying combinations and permutations of different components. Recently, there has been a shift towards parallel synthesis, primarily due to problems with deconvolution of complex combinatorial mixtures. These technologies still require classical thermal heat. The use of microwave chemistry in organic synthesis has now introduced a completely new approach to drug discovery. Microwave systems provide the opportunity to complete reactions in minutes, offering the option to return to more sequential methods.

As microwave synthesis instrumentation continues to evolve, new applications will be developed for a variety of chemistries and process developing needs.

This is advantageous because it allows chemists to analyze a reaction before conducting the next step, enabling them to optimize their reactions and their time. This chapter will document the many synthetic applications that have benefited from the use of microwave irradiation. *Note: The reader should assume that all reaction schemes shown in this chapter utilize microwave irradiation. In multi-step schemes, the use of microwave energy is indicated by "microwaves" on the arrow.*

A majority of the applications found in this chapter have been performed in a multi-mode microwave cavity under atmospheric conditions or in sealed glass or Teflon™ vessels, as single-mode reactors may not have been available at the time the work was completed. The rates of reactions performed in a multi-mode cavity are greater than those using conventional methods, but repeatability is low. The reader should also be aware that multi-mode instruments require a lot of power because of their spacious cavity. The total power generated is high, but the power density in the cavity is quite low. A higher power density allows the energy to be more focused in single-mode instrumentation, and 300 W or lower is sufficient. A chemist looking to mimic the conditions found in the references should not concentrate on the power level.

Chemists have been conducting research in microwave synthesis since the mid-1980s. As a result, there are many articles on the multitude of reactions that can be performed with microwave energy. Older reviews on microwave-enhanced synthetic applications include those by Abramovitch[296], Caddick[382], Majetich and co-workers[10,223,182], Sridar[294], and a more recent review by Lidstrom et al.[183] As microwave synthesis continues to grow in popularity, the applications written for it will multiply as well, though there are many types of reactions investigated in current literature including organometallic, cycloaddition, heterocyclic, oxidations, and condensations.

I. Organometallic cross-coupling reactions

Reactions that form carbon–carbon bonds are of supreme importance in synthetic chemistry. Palladium catalyzed cross-coupling reactions have become a significant part of drug discovery. Heck, Suzuki, and Stille coupling reactions are easily performed with microwave synthesis instrumentation. It was first suggested that microwave irradiation could enhance Suzuki and Stille cross-coupling reactions in June 1988 by Mills and co-workers at Glaxo in London, England.[184,185] Wali et al. performed the first Heck reaction in a multi-mode cavity in 1995.[186] He reacted iodobenzene with 1-decene (Scheme 33) and was able to get a complete reaction in approximately ten minutes compared to 14 hours with conventional methods.

Scheme 33

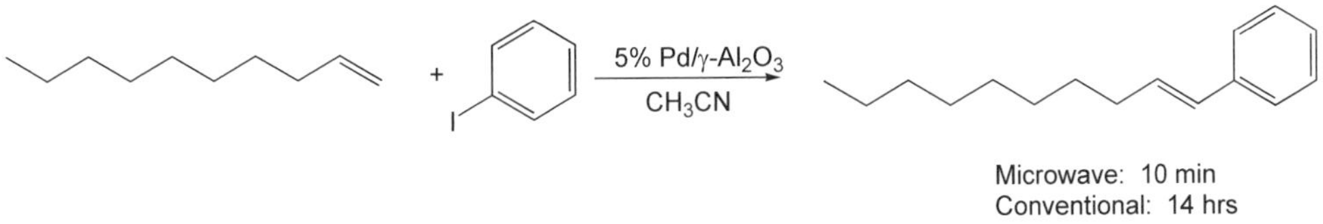

5% Pd/γ-Al_2O_3, CH_3CN

Microwave: 10 min
Conventional: 14 hrs

The second microwave work in palladium chemistry was done by Villemin and co-workers in France and presented in 1995 at a conference in Spain.[187] His palladium catalyzed Heck reaction (Scheme 34) was the first reported work in a single-mode cavity. Like Wali, his work yielded results in about 10 minutes, using only 140 W of power.

Scheme 34

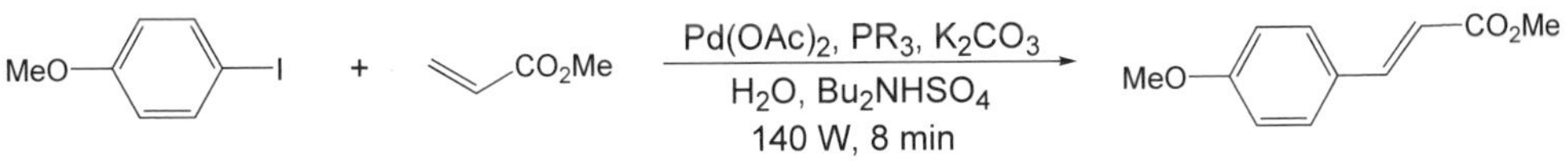

MeO–C_6H_4–I + CO_2Me (methyl acrylate) → MeO–C_6H_4–CH=CH–CO_2Me

$Pd(OAc)_2$, PR_3, K_2CO_3
H_2O, Bu_2NHSO_4
140 W, 8 min

More recently, Larhed and Hallberg, from the University of Upsalla in Sweden, have also explored palladium-coupling Heck reactions (Scheme 35), as well as Suzuki (Scheme 36) and Stille couplings (Scheme 37).[188,189] Their work, which has been quite extensive, again shows the major advantages in using microwave energy — rapid reaction times and increased yields.[13,188-201]

Scheme 35

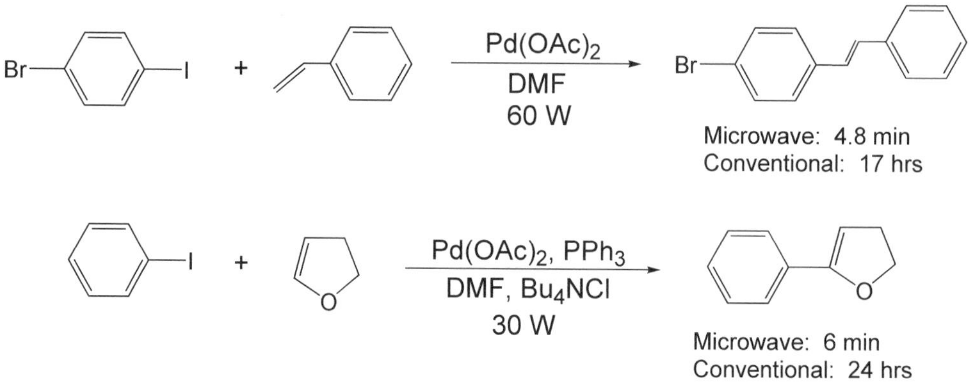

Br— —I + → Br—

$Pd(OAc)_2$
DMF
60 W

Microwave: 4.8 min
Conventional: 17 hrs

—I + O →

$Pd(OAc)_2$, PPh_3
DMF, Bu_4NCl
30 W

Microwave: 6 min
Conventional: 24 hrs

Scheme 36

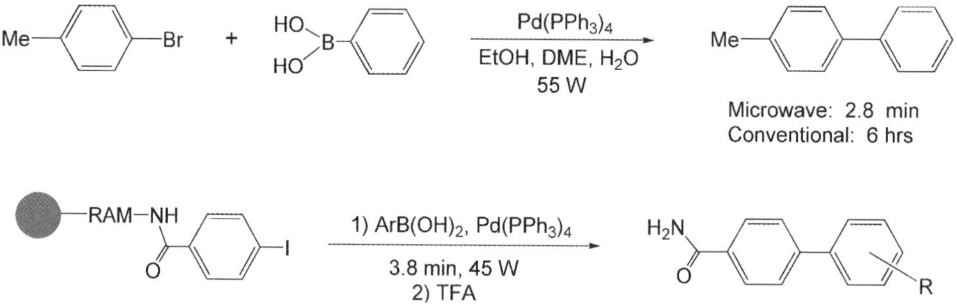

Me— —Br + HO, HO B— → Me—

$Pd(PPh_3)_4$
EtOH, DME, H_2O
55 W

Microwave: 2.8 min
Conventional: 6 hrs

—RAM—NH, O, —I → H_2N, O, R

1) $ArB(OH)_2$, $Pd(PPh_3)_4$
3.8 min, 45 W
2) TFA

Scheme 37

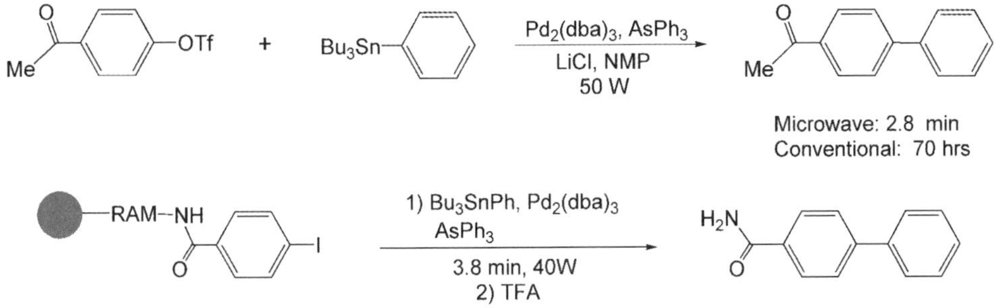
O
Me
OTf
+
Bu3Sn
Pd2(dba)3, AsPh3
LiCl, NMP
50 W
O
Me
Microwave: 2.8 min
Conventional: 70 hrs
RAM–NH
O
I
1) Bu3SnPh, Pd2(dba)3
AsPh3
3.8 min, 40W
2) TFA
H2N
O

Other research groups have been working in the areas of heteroaromatic synthesis and aqueous or solvent-free conditions via transition metal cross-coupling reactions.[99,117,202-207] Villemin and co-workers have executed Suzuki reactions in both water and solvent free conditions (Scheme 38).[206,207] Alternatively, Combs et al. have performed Suzuki-like reactions using a copper catalyst (Scheme 39).[204]

Scheme 38

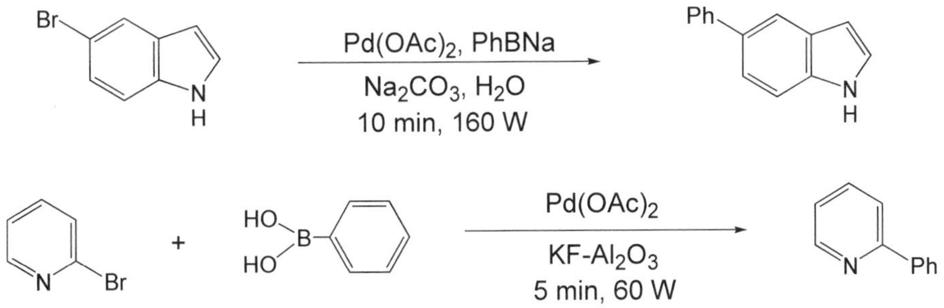
Br
N
H
Pd(OAc)2, PhBNa
Na2CO3, H2O
10 min, 160 W
Ph
N
H
N
Br
+
HO
B
HO
Pd(OAc)2
KF-Al2O3
5 min, 60 W
N
Ph

Scheme 39

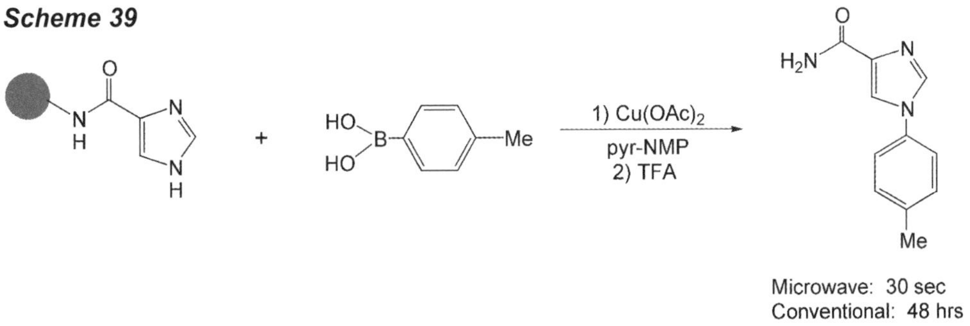
O
N
H
N
N
H
+
HO
B
HO
Me
1) Cu(OAc)2
pyr-NMP
2) TFA
O
H2N
N
N
Me
Microwave: 30 sec
Conventional: 48 hrs

Another area that has been recently explored is microwave-assisted Negishi cross-coupling reactions. The majority of the reactions discussed previously utilize aryl triflates, bromides, and iodides. Aryl and vinyl chlorides, unfortunately, have been quite unsuccessful in coupling reactions. The carbon–chlorine bond has a much larger bond dissociation energy than the others, which makes it harder to break. The Negishi cross-coupling reaction employs organozinc reagents, and it is a powerful solution to this dilemma. It opens up the use of an entire family of chlorides that are inexpensive and commercially available. With conventional heating, the Negishi reaction can require hours of heating for completion. Scheme 40 shows a reaction with an aryl bromide, which was complete in one minute and in a 90% yield.[208] The ease of cross-coupling between aryl chloride derivatives and organozinc halides is exhibited in Scheme 41.[18]

Scheme 40

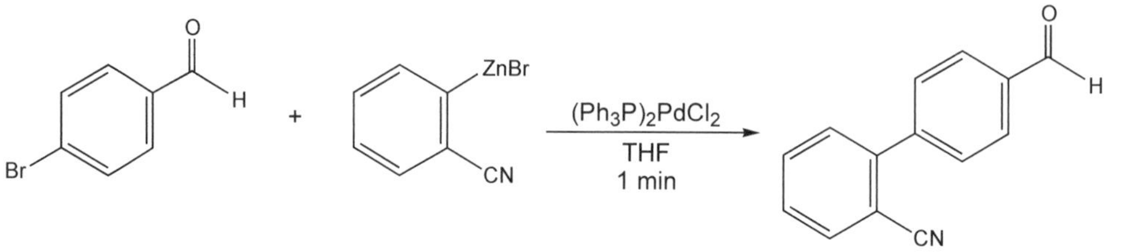

O H Br + ZnBr CN $(Ph_3P)_2PdCl_2$ THF 1 min O H CN

Scheme 41

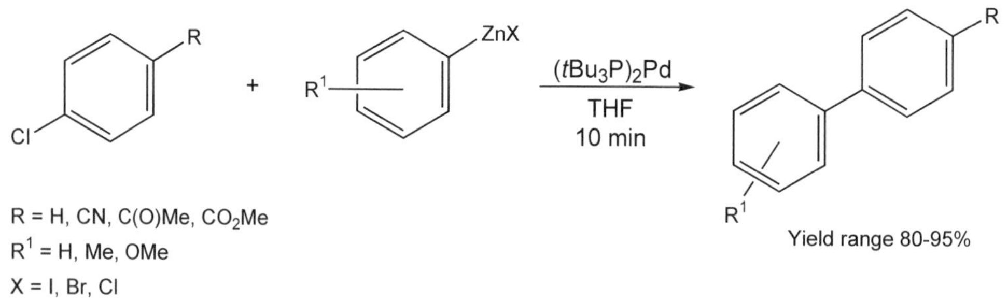

R Cl + ZnX R^1 $(tBu_3P)_2Pd$ THF 10 min R R^1

R = H, CN, C(O)Me, CO_2Me

R^1 = H, Me, OMe

X = I, Br, Cl

Yield range 80-95%

II. Cycloadditions

Cycloadditions, which include the Diels-Alder, ene, and Alder-Bong reactions, are important single-step, ring-forming reactions in organic synthesis. These transformations usually require harsh conditions (high pressures and temperatures) and long reaction times. Diels-Alder cycloadditions were the first reaction types to be examined in conjunction with microwave irradiation.[209-236,262] Giguere et al. showed one of the first examples of a microwave-induced Diels-Alder reaction in 1986 (Scheme 42).[3] Irradiating anthracene and dimethyl fumarate in a multi-mode instrument at 600 W gave a complete reaction in 10 minutes compared to 72 hours with conventional heating.

Scheme 42

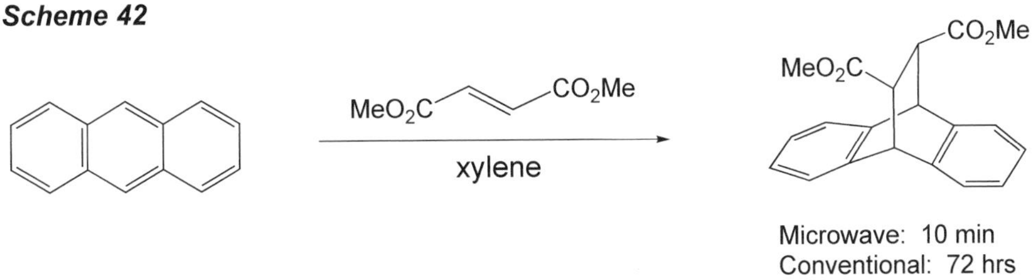

MeO_2C / CO_2Me (dimethyl fumarate), xylene

MeO_2C, CO_2Me

Microwave: 10 min
Conventional: 72 hrs

Majetich and Hicks have also shown successful Diels-Alder reactions.[223] Both of the transformations shown in Schemes 43 and 44 were executed in DMF or in solvent-free conditions.

Scheme 43

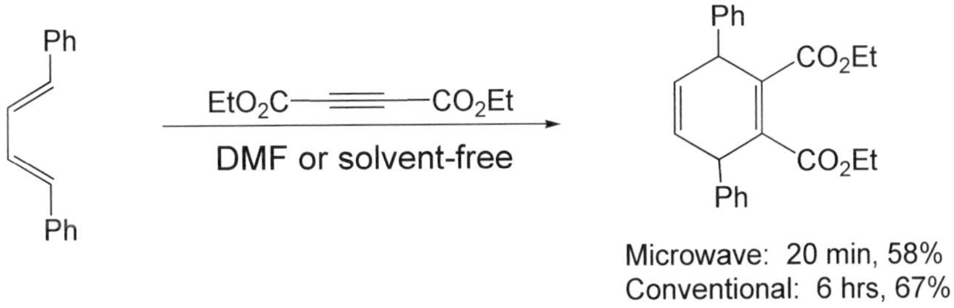

Ph, Ph

EtO_2C—≡—CO_2Et

DMF or solvent-free

Ph, CO_2Et, CO_2Et, Ph

Microwave: 20 min, 58%
Conventional: 6 hrs, 67%

Scheme 44

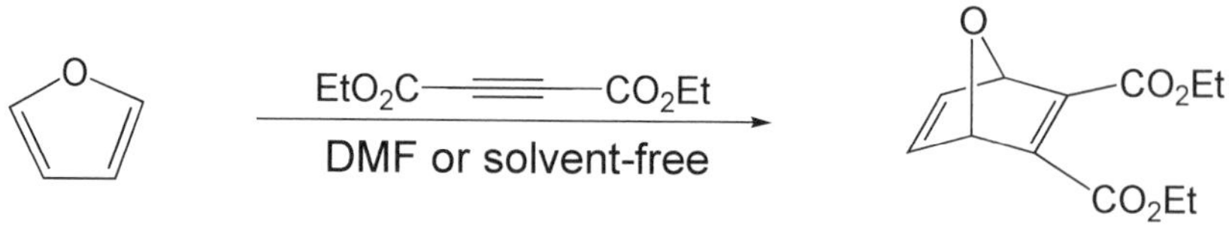

EtO_2C—≡—CO_2Et
DMF or solvent-free

CO_2Et
CO_2Et

Microwave: 10 min, 86%
Conventional: 15 min, 68%

Microwave heating has also been used extensively in heterocyclic Diels-Alder reactions.[224-236,263] These reactions are very important in synthetic chemistry, as they enable the synthesis of biologically significant nitrogen-, sulfur-, and oxygen-containing rings, which are usually difficult to achieve by standard methodology. Avalos et al. reacted 1,2-diaza-1,3-butadienes with diethyl azodicarboxylate (DEAD) to form functionalized tetrazines (Scheme 45).[224] Under conventional heating, these reactions take 30 days for completion, whereas with microwave irradiation, they were performed in 15 minutes.

Scheme 45

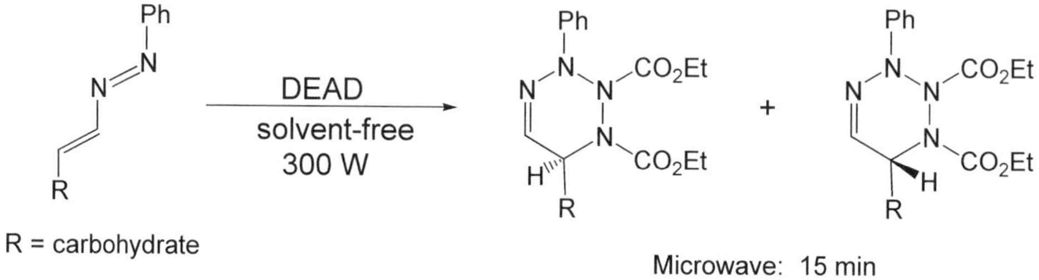

Ph
N
N
R
R = carbohydrate

DEAD
solvent-free
300 W

Ph N N N N CO_2Et CO_2Et H R
+
Ph N N N N CO_2Et CO_2Et H R

Microwave: 15 min
Conventional: 30 days

The ene reaction is a reaction between and alkene and an enophile (analogous to a dienophile in a Diels-Alder reaction). A new C–C bond is formed, and the position of the original double bond shifts through a cyclic transition state. Intramolecular reactions are entropically more favorable, but they still require long reaction times to complete. One example of an intramolecular ene reaction is shown in Scheme 46.[3] With microwave irradiation, this solvent-free reaction only took 15 minutes compared to 12 hours with conventional methods.

Scheme 46

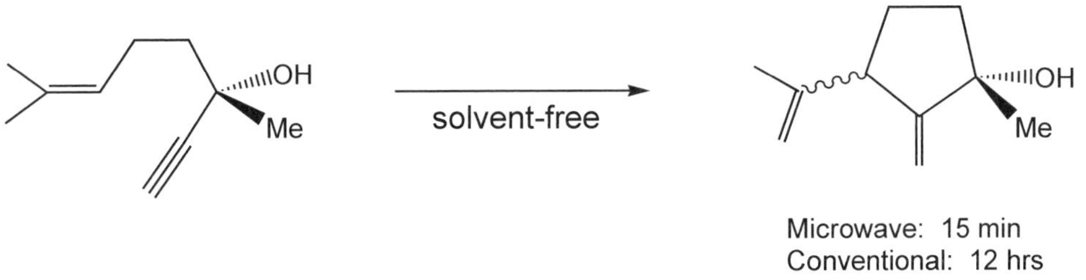

OH
Me
solvent-free
OH
Me

Microwave: 15 min
Conventional: 12 hrs

The Alder-Bong reaction is another interesting cycloaddition reaction. In this particular reaction, an ene reaction is followed by an intramolecular Diels-Alder reaction. The Diels-Alder reaction proceeds rapidly, thus the intermediates formed from the ene reaction cannot be isolated. Scheme 47 shows a reaction

Scheme 47

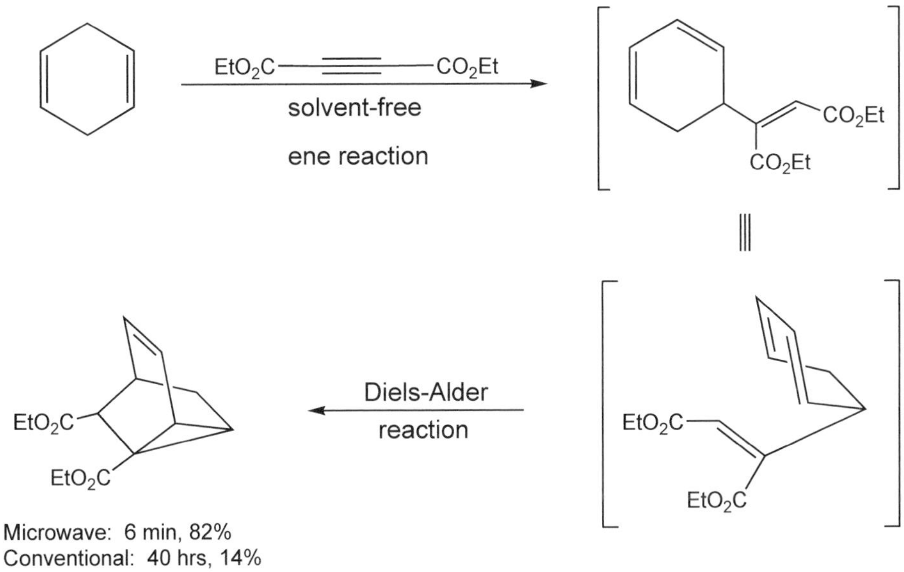

EtO_2C—≡—CO_2Et
solvent-free
ene reaction
CO_2Et
CO_2Et

Diels-Alder
reaction
EtO_2C
EtO_2C
EtO_2C
EtO_2C

Microwave: 6 min, 82%
Conventional: 40 hrs, 14%

between 1,4-cyclohexadiene and diethyl acetylene dicarboxylate.[237] With conventional heating methods, this reaction takes 40 hours to complete. With microwave irradiation, it proceeds in 20 minutes, solvent-free.

III. Heterocycles

Heterocyclic chemistry is another area of importance in synthetic chemistry. A large number of natural products and target drug compounds contain a heterocyclic core. Synthetic routes toward these compounds are usually quite challenging. Microwave-induced heterocyclic chemistry has been extensively examined with pyrimidine derivatives[49,125,130,131,169,238-251], pyrroles[252-257], pyridines[59,64,68,108,258-274], β-lactams[11,111,275-286], indoles[287-298], γ-carbolines[299], quinolines and quinolones[47,96,110,112,113,118,300-303], quinazolines[67,304-309], imidazoles[133,155-156,258,264,265,310-327], other azole/azoline derivatives[57,58,64,71,106,107,109,119,133,327-348], furans[349-353], and 1,3-dipolar cycloadducts[61-63,155,157,177,355-378].

The Biginelli three-component condensation reaction is a one-pot synthesis to dihydropyrimidines. These heterocyclic systems contribute to enhanced pharmacological efficiency in a variety of biological effects, including antiviral, antitumor, antibacterial, and anti-inflammatory activities. With normal conventional heating, these reactions can take approximately 24 hours

Scheme 48

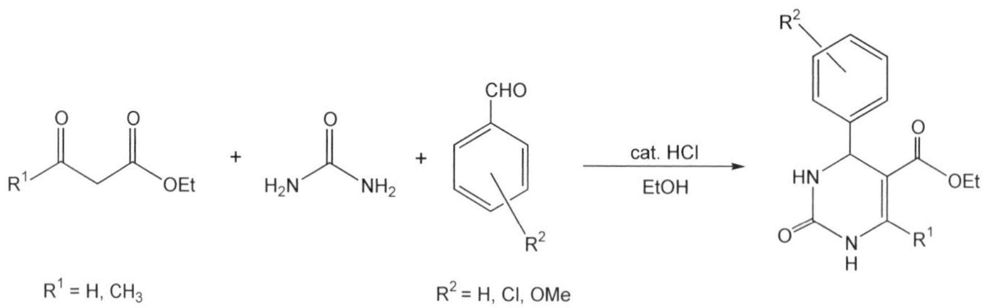

cat. HCl, EtOH

R^1 = H, CH_3

R^2 = H, Cl, OMe

Microwave: 5 min, 60-90% yield
Conventional: 12-24 hours, 15-60% yield

for complete transformation with only low to moderate yields. Upon microwave irradiation, the Biginelli reaction was successfully completed in five minutes, with 60-90% yields (Scheme 48).[354]

The Paal-Knorr condensation/cyclization reacts 1,4-diketones with primary amines to form N-substituted pyrroles. This synthesis requires at least twelve hours of prolonged thermal heating and added Lewis acids to activate the diketones. With microwaves, transformation occurred in anywhere from 30 seconds to two minutes with very high yields (75-90%) (Scheme 49).[255]

Scheme 49

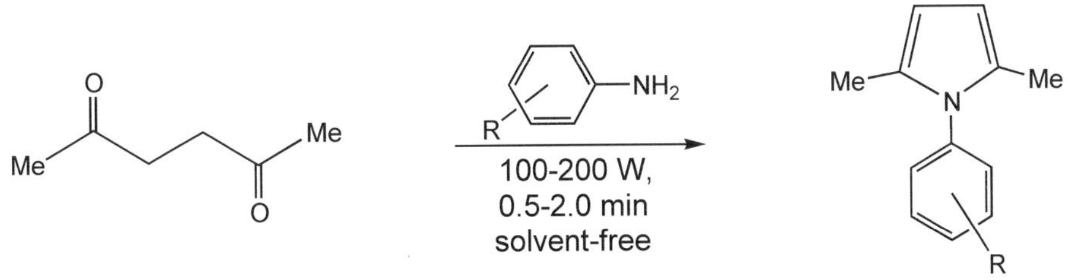

O, Me, Me, O

NH_2, R

100-200 W,
0.5-2.0 min
solvent-free

Me, Me, N, R

Substituted dihydropyridines are known to be calcium channel blockers and are quite biologically active. They can be synthesized via the one-pot Hantzsch pyridine reaction. In this particular reaction, an aldehyde, two equivalents of a β-ketoester, and ammonium hydroxide are combined in the same reaction vessel. One equivalent of the β-ketoester and the aldehyde undergo an aldol condensation. The other equivalent reacts with the ammonium hydroxide to yield an enamine. The final transformation, as shown in Scheme 50, results in a dihydropyridine derivative. Classical thermal heating takes over 24 hours, whereas these reactions occur in five minutes or less with microwave irradiation.[259,260,267]

Scheme 50

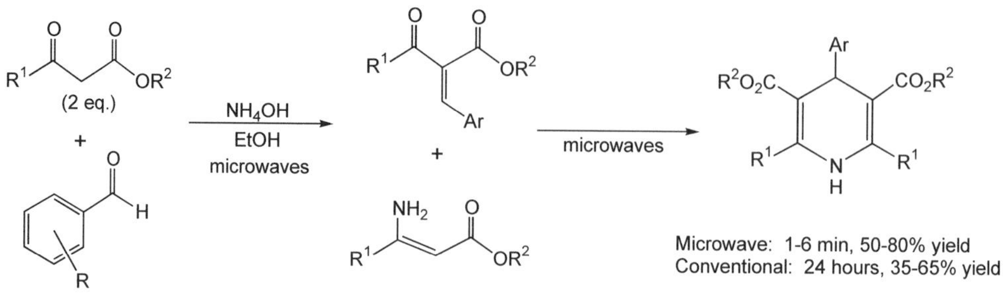

Microwave: 1-6 min, 50-80% yield
Conventional: 24 hours, 35-65% yield

For decades, synthetic and medicinal chemists have been greatly interested in β-lactams. These four membered chiral heterocycles are versatile synthons in natural product synthesis. Additionally, they can easily undergo rearrangements to yield other heterocyclic and acyclic compounds. Bose and coworkers have done extensive research on microwave-induced β-lactam synthesis.[11,278,279,282,283,285] Their early work on β-lactams via conductive heating led to extremely low yields. Utilization of microwave heating for five minutes on an α,β-unsaturated acid chloride and a Schiff base (imine) provides β-lactams in 65-70% yield (Scheme 51). Another synthetic strategy to β-lactams utilizes diazoketones (Scheme 52).[275] Upon irradiation with microwaves, the diazoketone compound is transformed into a ketene, which then rapidly cyclizes with the imine to yield a β-lactam. Classical conditions rarely yielded a product, whereas microwave irradiation produced 60-80% yield of β-lactam.

Scheme 51

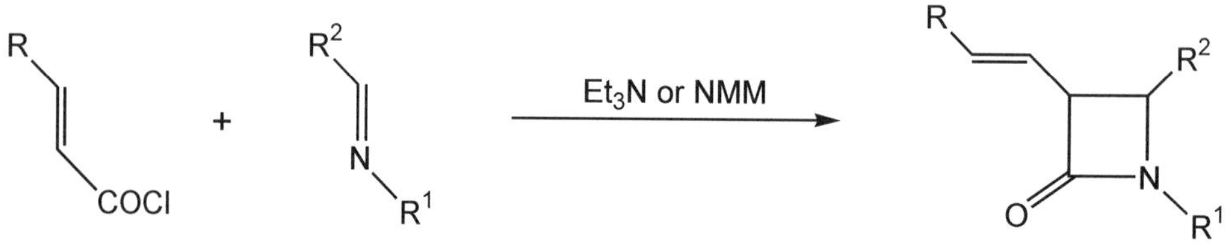

Et_3N or NMM

Scheme 52

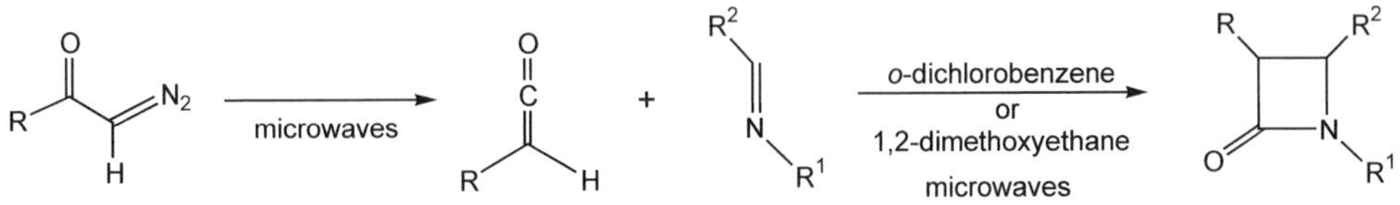

o-dichlorobenzene or 1,2-dimethoxyethane microwaves

The Fisher indole cyclization is a one-pot reaction to substituted indolic compounds. This powerful synthetic reaction utilizes an arylhydrazone and an aldehyde or ketone. After a [3,3]-sigmatropic rearrangement, elimination of ammonia yields an indole. Microwave irradiation greatly accelerates (385-fold rate enhancement) this rearrangement, resulting in successful completion of the reaction in less than 30 seconds (Scheme 53).[294-296]

Scheme 53

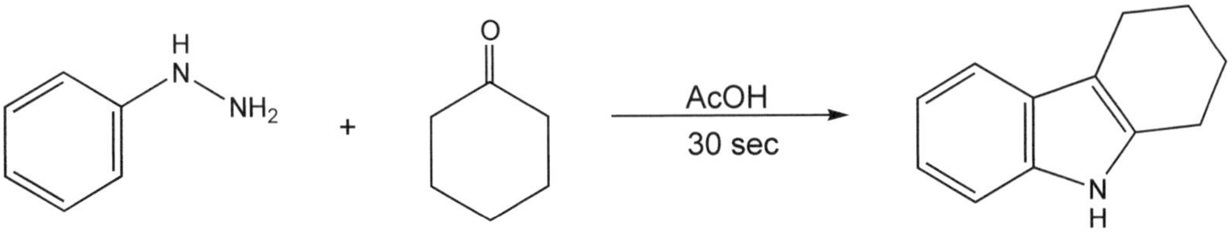

AcOH 30 sec

The Graebe-Ullmann synthesis, which yields γ-carbolines, is another one-pot reaction. This two-step procedure firsts reacts benzotriazole with a 4-chloropyridine followed by polyphosphoric acid addition to produce thermolytic ring closure. Traditionally, both steps require extremely high temperatures for reaction to occur, and the yields are very dependent on temperature control. With microwave heating, the first step occurs in ten minutes (160 W), plus an additional five minutes (or until nitrogen evolution ceases), after the addition of polyphosphoric acid (Scheme 54).[299]

Scheme 54

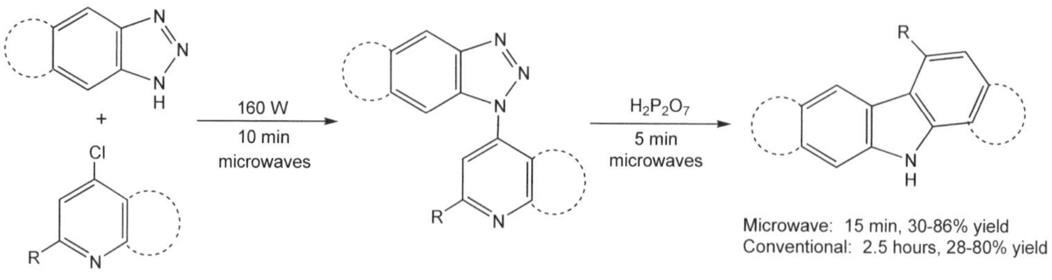

160 W
10 min
microwaves

$H_2P_2O_7$
5 min
microwaves

Microwave: 15 min, 30-86% yield
Conventional: 2.5 hours, 28-80% yield

Quinolines and quinazolines are hetero-bicyclic ring systems containing one and two nitrogen atoms, respectively. These compounds and their derivatives have been shown significant interest by medicinal chemists because of the numerous natural products and potential drug compounds that contain their heterocyclic core. The classic Skraup quinoline synthesis requires large amounts of sulfuric acid and high temperatures, which can be quite a violent combination, and usually does not produce satisfactory yields. With microwave irradiation, it has been reported that an aniline, an alkyl vinyl ketone, and indium(III) chloride (catalyst) on silica gel provide 4-alkylquinolines in 80-90% yield (Scheme 55).[300] Quinazolines can be synthesized from an N-arylimino dithiazole derivative and sodium hydride in refluxing ethanol. Reaction times were greatly reduced from 40 hours to two hours with microwave energy (Scheme 56).[304]

Scheme 55

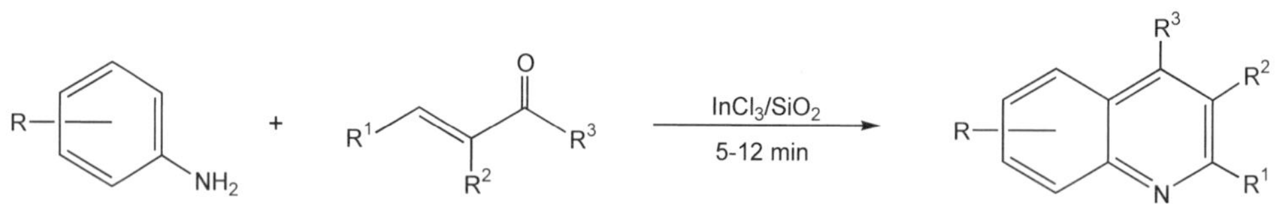

$InCl_3/SiO_2$
5-12 min

Scheme 56

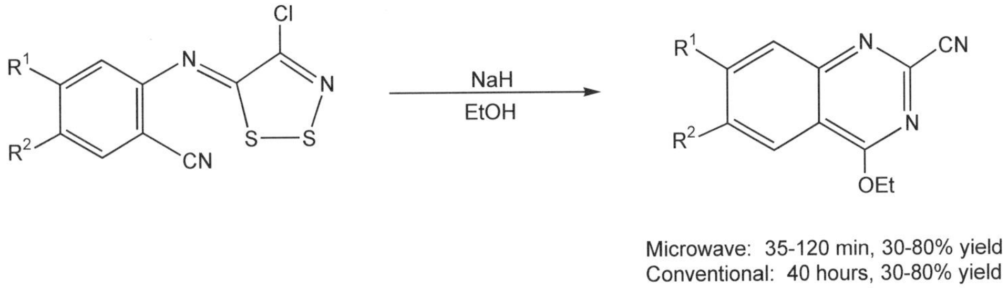
Cl
R1
R2
N
N
S
S
CN
NaH
EtOH
N
CN
N
OEt
Microwave: 35-120 min, 30-80% yield
Conventional: 40 hours, 30-80% yield

Azole derivatives — which include imidazoles, oxazoles, thiazoles, and tri/tetrazoles — are five-membered rings containing at least two heteroatoms, one being nitrogen. Once again, these compounds contain very important heterocyclic cores that are common in drugs and natural products. Benzimidazoles, -oxazoles, and -thiazoles can be easily synthesized from *o*-arylenediamines or other *o*-arylene heteronucleophiles and a substituted acetylketene diethyl acetal with microwave irradiation (Scheme 57).[327] No reaction occurred with thermal heat.

Scheme 57

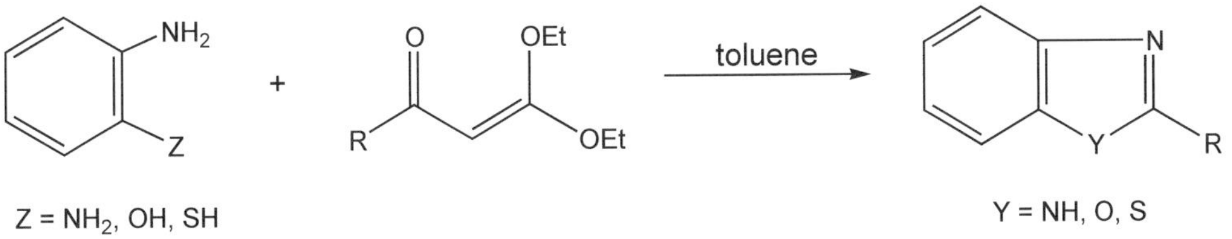
NH2
Z
+
O
OEt
R
OEt
toluene
N
Y
R
Z = NH2, OH, SH
Y = NH, O, S

Another route to substituted imidazoles can be accomplished by condensation of a 1,2-dicarbonyl compound with an aldehyde and ammonia. Classically, this reaction requires four hours of intense heating in acetic acid. With microwave irradiation, imidazoles were produced in 20 minutes (Scheme 58).[316]

Scheme 58

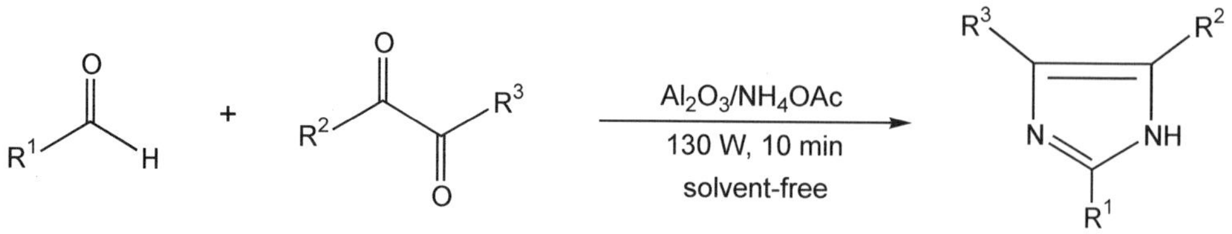
O
R1
H
+
O
R2
R3
O
Al2O3/NH4OAc
130 W, 10 min
solvent-free
R3
R2
N
NH
R1

Triazoles and tetrazoles contain, as their prefixes indicate, three and four nitrogen atoms, respectively. Treatment of arylnitriles with hydrazine dichloride and hydrazine hydrate in ethylene glycol provides substituted 1,2,4-triazoles. Conventional methods require 45-60 minutes of intense heating and provide moderate yields. Microwave irradiation produced the triazole products in about five minutes (Scheme 59).[345] Tetrazoles can also be synthesized from nitriles, these via palladium-catalyzed cyanation of organo bromides with zinc cyanide. Subsequent addition of sodium azide and ammonium chloride yield tetrazoles. Conductive heating usually takes from seven hours to four days, while microwaves yielded product in 15 minutes (Scheme 60).[348]

Scheme 59

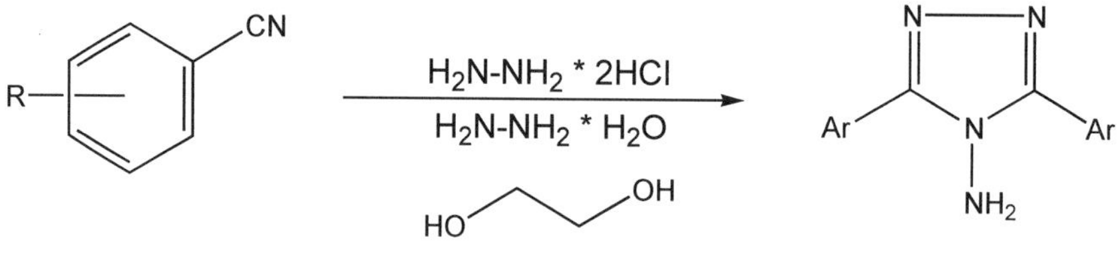
CN
R
H2N-NH2 * 2HCl
H2N-NH2 * H2O
HO
OH
N
N
Ar
N
Ar
NH2

Scheme 60

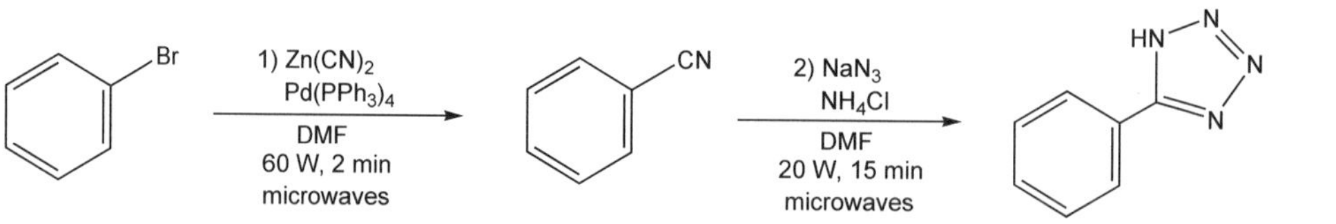
Br
1) Zn(CN)2
Pd(PPh3)4
DMF
60 W, 2 min
microwaves
CN
2) NaN3
NH4Cl
DMF
20 W, 15 min
microwaves
HN
N
N
N

Microwave: 10-15 min, 50-98%
Conventional: 7-96 hours, 35-97%

Furans also contain a common heterocyclic core that is seen in many natural products and drug compounds. Diepoxides, in the presence of sodium iodide, rearrange to form substituted furan ethers. Using thermal heat, this reaction proceeds in five hours with only a 43% yield. With microwave irradiation, rearrangement only took five minutes with an 88% product yield (Scheme 61).[352] Additionally, ferrocenyl substituted acrylaldehydes react with a β-hydroxy (or thiol) ester to yield furans (thiophenes) in two minutes under microwave heating (Scheme 62).[353] Conventional methods require 24 hours of reflux.

Scheme 61

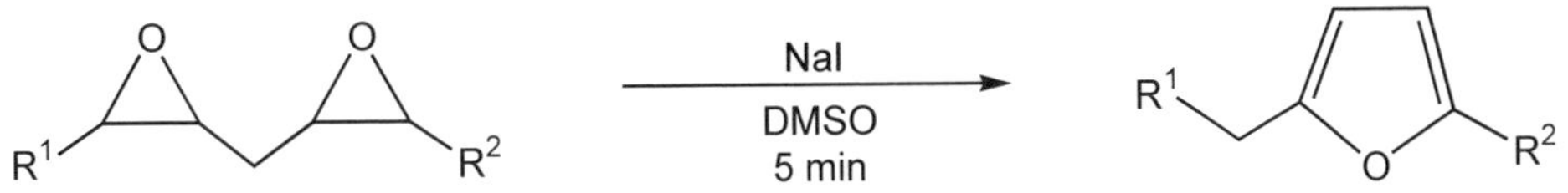

NaI
DMSO
5 min

Scheme 62

1:2 Et_3N:DMF
2 min

X = O, S

X = O, S

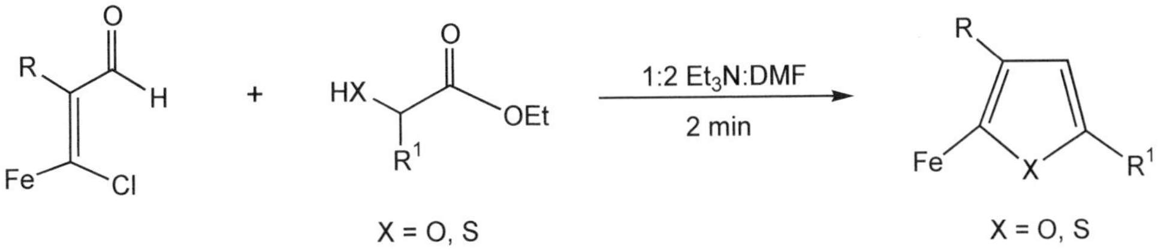

A 1,3,4-oxadiazole synthesis, as shown in Scheme 63, was executed in two steps.[354] The second stage of this reaction was performed, first, with conventional heating methods at 150 °C for 90 minutes. In a separate reaction using microwave energy, the second stage was completed at the same temperature, but it only took five to ten minutes. This reaction was 100-fold faster, even though the measured bulk temperatures were the same.

Scheme 63

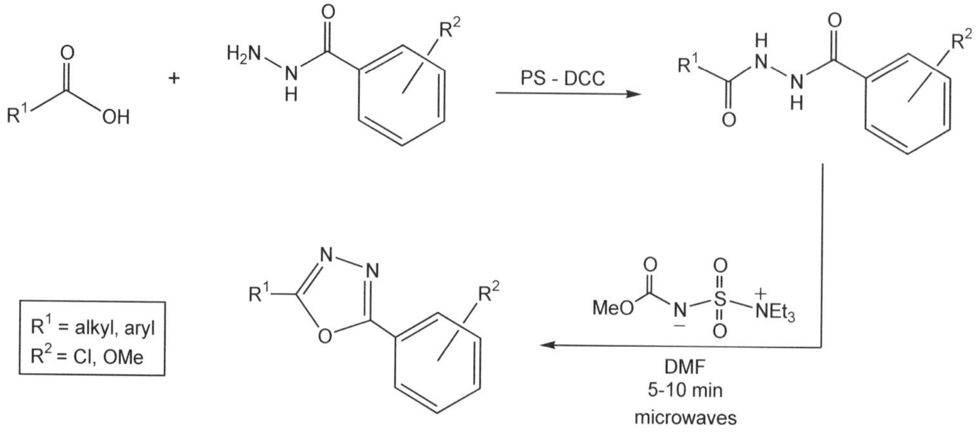
R1
OH
O
+
H2N
N
H
O
R2
PS - DCC
R1
H
N
N
H
O
O
R2
MeO
N
S
O
O
O
NEt3
DMF
5-10 min
microwaves
R1
N
N
O
R2
R1 = alkyl, aryl
R2 = Cl, OMe

Another organic transformation that benefits from the use of microwave irradiation is a 1,3-dipolar reaction.[23,61-63,155,157,177,355-378] These types of reactions are cycloadditions, but they are being discussed in this section because they form heterocyclic moieties. One of the main reactants, which are generated in situ, is a 1,3-dipole. 1,3-Dipoles are heteroatom molecules that contain both a positive and a negative charge. Due to ionic conduction, these ionic species will directly interact with the microwave energy being applied. They readily react with a dipolarophile, which is usually an electrophilic alkene or alkyne, to form heterocyclic ring systems. Using conventional methods, these cycloadditions can take 1-2 days. In Scheme 64, nitrile oxides (X = O) and nitrile imines (X = NPh) are irradiated with microwaves to yield different heterocyclic compounds in only 30 minutes.[355]

Scheme 64

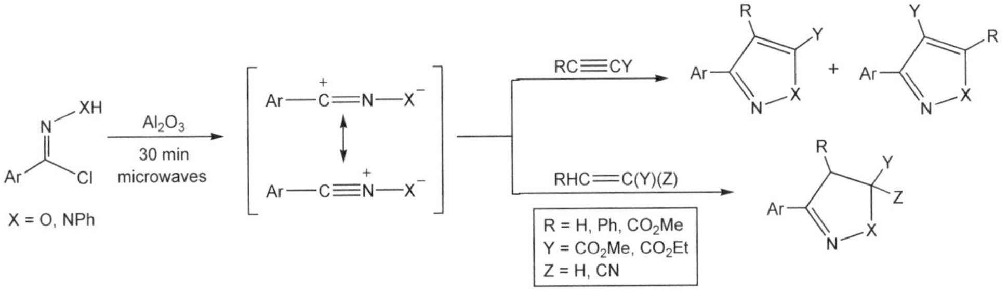

IV. Nucleophilic additions and substitutions

Nucleophilic addition and substitution reactions, both aromatic and aliphatic, encompass a large number of synthetic transformations. Microwave irradiation has been used extensively to enhance nucleophilic aromatic substitutions[40,110,379-387], Michael additions[148,192,388-402], Mitsunobu reactions[403], hydroacylations[404], N-acylations[41,42,90,405-422], acetylations[422-424], carbon[382,389,425-439] and heteroatom alkylations (N[66,172,265,317,421,439-464], S[465-469], O[53-54,134,179,351,380,425,464,470-485]), Williamson etherifications[4,10,11,223,486-491], esterifications[56,423,492-506], transesterifications[41,158,507,508], halogenations[10,181,223,510-518], and ^{18}F-radiolabeling[519,520].

Nucleophilic aromatic substitution (S_NAr) reactions play an important role in drug discovery. A large number of drug compounds contain multiple aromatic rings. S_NAr allows an organic chemist a facile route to changing substituents on the ring systems. Classically, S_NAr requires long reaction times and high temperatures and provide low to moderate product yields. Scheme 65 shows two different substitutions on 1-chloro-4-nitrobenzene. Route **A** shows a substitution to an amine with ammonia and copper(I) oxide.[381] With microwave irradiation, this transformation is successfully completed in one hour with a 93% product yield. Likewise, in route **B**, replacing the chlorine substituent with an ethoxy group forms an ether quantitatively in only two minutes.[382]

Scheme 65

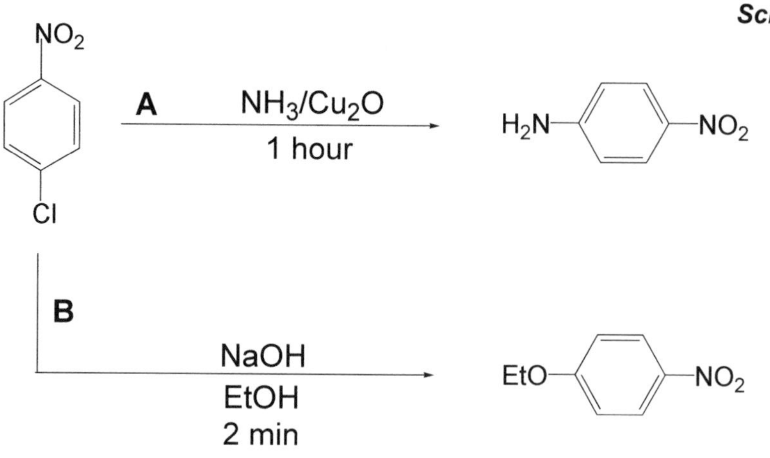

NO_2

Cl

A NH_3/Cu_2O 1 hour

H_2N—NO_2

B NaOH EtOH 2 min

EtO—NO_2

Scheme 66 exhibits a small library of heterocyclic compounds that were synthesized using S_NAr.[354] Starting from one common aromatic scaffold; different amines were added, individually, to yield a small family of eight compounds. Using microwave instrumentation, this entire library was achieved in less than 90 minutes, whereas with conventional methods, this could take many days to complete. Additionally, the yields of this reaction greatly increased from as high as 60% with conventional heating to quantitative yields with microwave irradiation.

The Michael reaction forms the basis for many synthetic transformations. It involves a conjugate 1,4-addition of a nucleophile to an α,β-unsaturated ketone, aldehyde, amide, nitrile, nitride, sulfoxide, or sulfone. Scheme 67 shows an example of a Michael addition reaction between two indoles.[388] Bis(indole) molecules have recently been isolated from sponges and are known bioactive metabolites. In this particular reaction, both the nitrovinylindole and alkylindole are adsorbed on silica gel and then subjected to microwave irradiation for 7-10 minutes. Using conventional methods, these additions proceeded in considerably longer reaction times, 8-14 hours.

Scheme 66

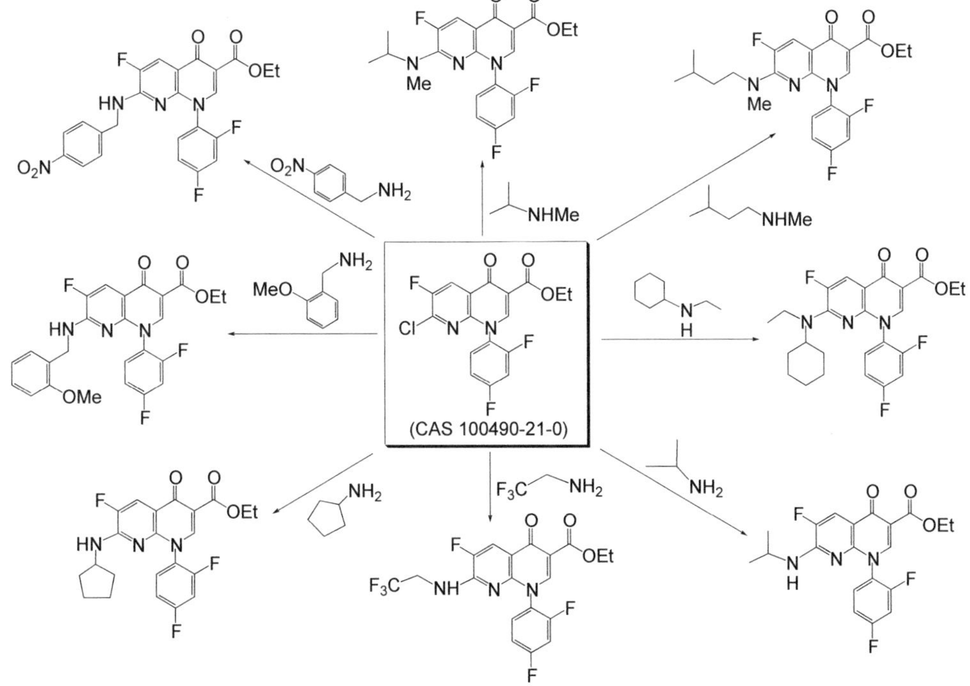

(CAS 100490-21-0)

Microwave: 1.2 eq. amine, MeCN, 175°C, 10 min. All reactions afforded quantitative yields, based on LCMS.
Conventional: 100°C, 12-24 hours, 35-60% yield

Scheme 67

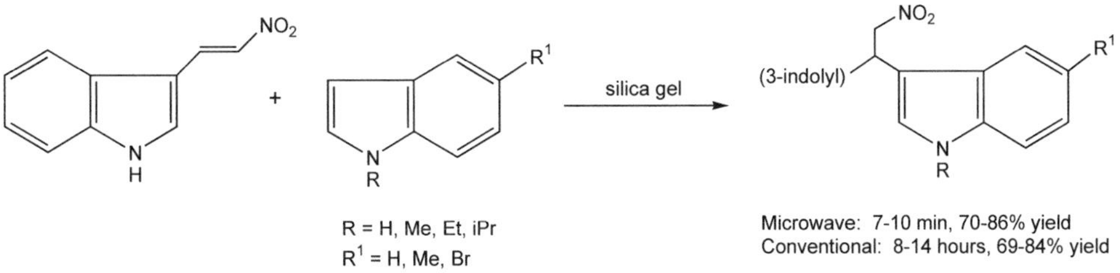

silica gel

R = H, Me, Et, iPr
R^1 = H, Me, Br

Microwave: 7-10 min, 70-86% yield
Conventional: 8-14 hours, 69-84% yield

Mitsunobu reactions are powerful synthetic transformations that can invert stereochemical configurations. Microwave irradiation has been used to enhance these conversions, as classical methods usually require high temperatures and long reaction times. Scheme 68 exhibits the acetylation of (*S*)-sulcatol via microwave enhanced Mitsunobu conditions (triphenylphosphine/diisopropyl azodicarboxylate) with acetic acid followed by lithium aluminum hydride (LAH) reduction to (*R*)-sulcatol.[403]

Scheme 68

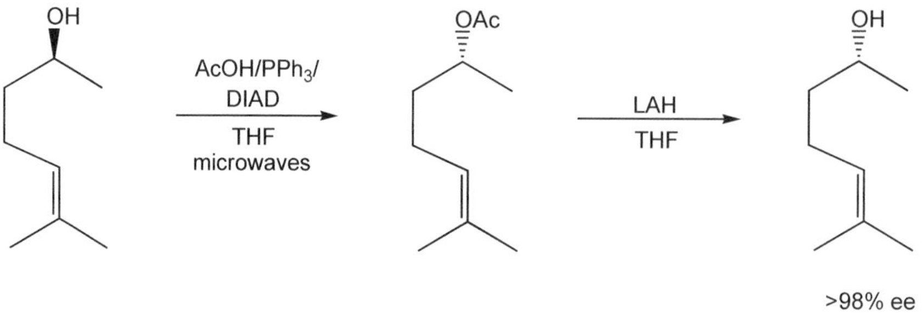
OH
AcOH/PPh3/
DIAD
THF
microwaves
OAc
LAH
THF
OH
>98% ee

Acylations [-C(O)R] and acetylations [-C(O)Me] are useful synthetic methods for obtaining ketones, amides, and enol esters. In hydroacylation reactions, aldehydes and olefins yield ketones via C–H bond activation by transition metals. Scheme 69 shows an efficient synthesis to ketones utilizing both Wilkinson's rhodium(I) catalyst and 2-amino-3-picoline.[5] The aldimine intermediate

Scheme 69

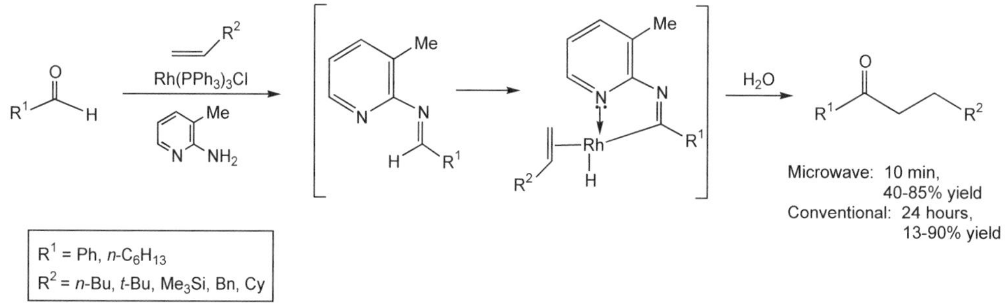
R2
O
R1
H
Rh(PPh3)3Cl
Me
N
NH2
Me
N
N
H
R1
Me
N
N
Rh
H
R1
R2
H2O
O
R1
R2
Microwave: 10 min, 40-85% yield
Conventional: 24 hours, 13-90% yield
R1 = Ph, n-C6H13
R2 = n-Bu, t-Bu, Me3Si, Bn, Cy

undergoes cyclometallation with the rhodium catalyst, which is then followed by alkene coordination. Alkene insertion followed by reductive elimination yields a ketone.[404] Thermal methods can take 24 hours, but those reaction times have been reduced to four hours with a benzoic acid catalyst. With microwave heating, the reaction proceeds in ten minutes with moderate to high product yields.

Conversion of amines to amides is the most widely used protection method for amino groups. N-acylation of amines to maleimides is useful and can be enhanced with microwave heating (Scheme 70).[405-410] Trifluoroacetylation is quite convenient in organic synthesis because of its facile cleavage. This acetylation is usually achieved with trifluoroacetic anhydride, but having trifluoroacetic acid as a byproduct causes one to look for alternative methods. The use of $TiO(CF_3CO_2)$ provides a solution, as titanium oxide and water are the only byproducts. The use of this reagent on both primary and secondary amines, coupled with 5-10 minutes of microwave irradiation, gives trifluoroacetamides in excellent yields (Schemes 71 and 72).[422] Use of conventional heating with the same reagents and reaction conditions takes at least 48 hours.

Scheme 70

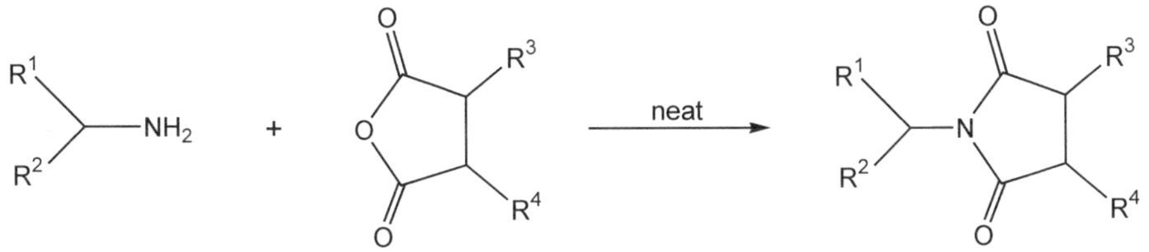

neat

Scheme 71

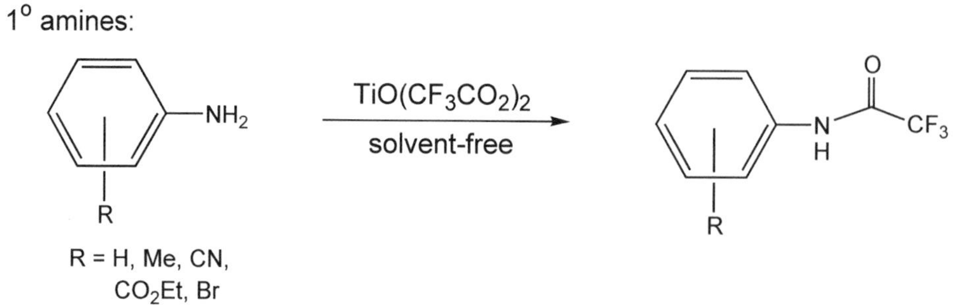

1° amines:

NH_2 — R — R = H, Me, CN, CO_2Et, Br

$TiO(CF_3CO_2)_2$, solvent-free →

N, H, O, CF_3, R

Scheme 72

2° amines:

NH, Ph_2NH

$TiO(CF_3CO_2)_2$, solvent-free →

$NCOCF_3$, Ph_2NCOCF_3

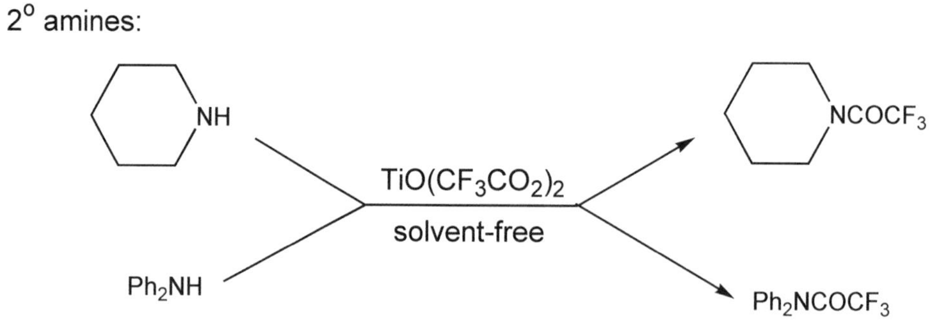

Enol-acetylation of ketones is a valuable transformation in organic chemistry. The enol ethers that are formed are used extensively as intermediates in synthetic routes. Despite their popular usage, preparation methods are limited. A common procedure involves acetic anhydride with a basic or acid catalyst. These catalysts are very strong and can cause sensitive compounds to decompose. Scheme 73 shows a mild procedure that selectively acetylates six-membered cyclic ketones.[423] With conductive heating, the cyclohexanone derivatives were refluxed in THF with acetic anhydride and iodine for 8 hours and gave very low product yields. Alternatively, quantitative yields were produced with microwave irradiation in only five to ten minutes.

Scheme 73

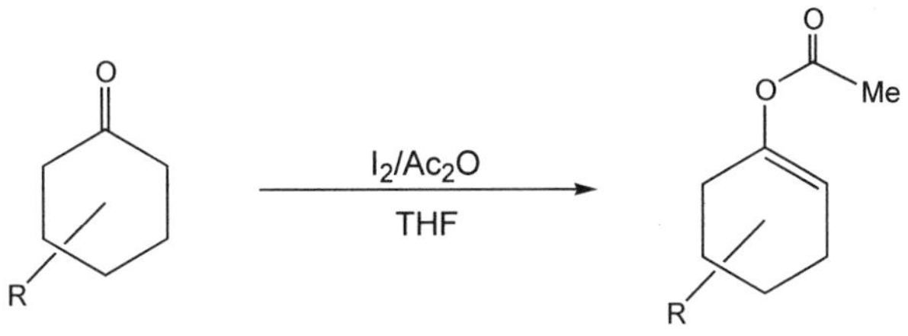
O
R
I_2/Ac_2O
THF
O
O
Me
R

Carbon-alkylations are also important reactions in organic chemistry. The most common and well-known method is by deprotonating a carbon that is adjacent to an electron-withdrawing group (e.g. enolate formation). Scheme 74 shows alkyl addition to a substituted acetate.[382] With microwave heating, this reaction proceeded in 3 minutes with high product yields. Another example of C-alkylation is the addition of a 2-nitropropane anion (**A**) with a heterocyclic electrophile (**B**) (Scheme 75). The solvent conditions determine whether the final product is **C** or **D**, which is formed from subsequent elimination of nitrous acid from **C**.[426]

Scheme 74

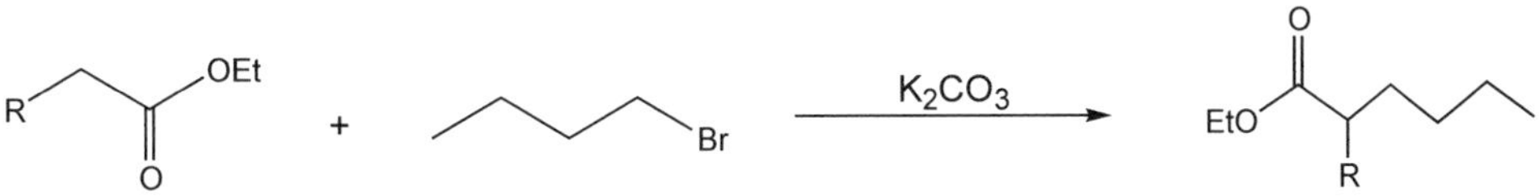
R
OEt
O
+
Br
K_2CO_3
O
EtO
R

Scheme 75

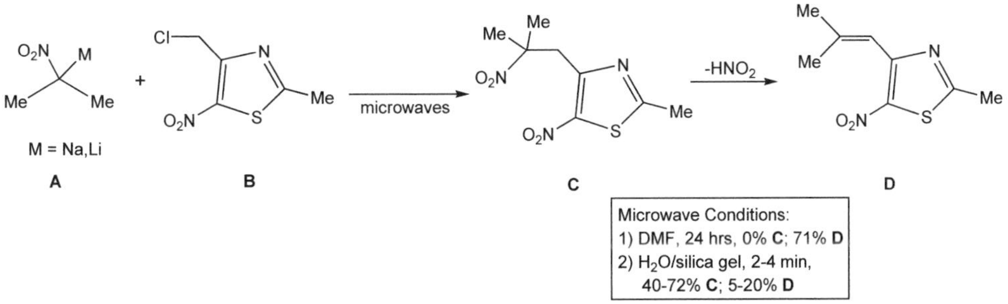
O_2N
M
Me
Me
M = Na,Li
A
+
Cl
N
Me
O_2N
S
B
microwaves
Me
Me
O_2N
N
Me
O_2N
S
C
$-HNO_2$
Me
Me
N
Me
O_2N
S
D
Microwave Conditions:
1) DMF, 24 hrs, 0% C; 71% D
2) H_2O/silica gel, 2-4 min,
40-72% C; 5-20% D

Heterocyclic alkylation reactions also benefit from microwave irradiation. Saccharin can easily be alkylated with any alkyl halide under microwaves in only ten minutes (Scheme 76).[445] The saccharin is first treated with base to form the sodium salt, which is then adsorbed on silica gel. This reaction is solvent-free and gives a 91% yield. Thiols can also be alkylated in near quantitative yield with alkyl halides via potassium carbonate on alumina (Scheme 77).[450]

Scheme 76

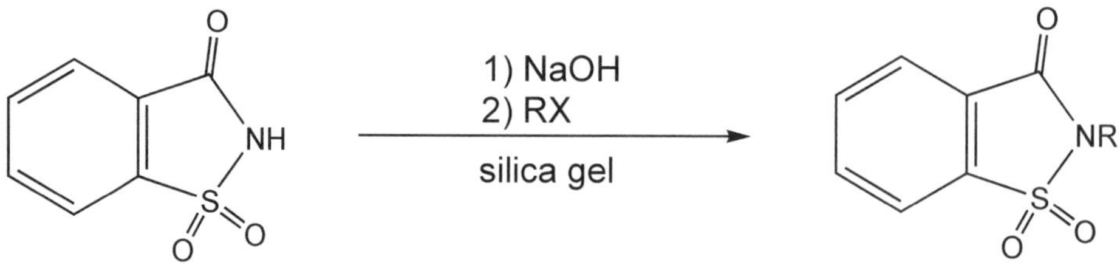

1) NaOH
2) RX
silica gel

Scheme 77

R^2—X
K_2CO_3/Al_2O_3

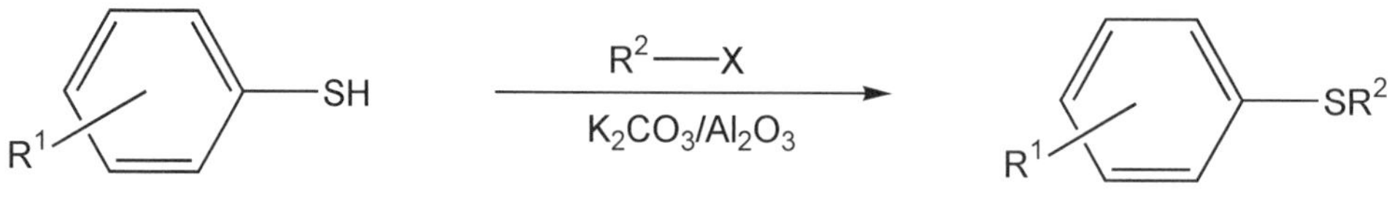

Oxygen-alkylation of phenolic compounds is a versatile approach to aryl ethers. These compounds are the basis of many pharmaceutical templates that are used in drug discovery. With conductive heating, these reactions can take anywhere from one to seven days for completion. Microwave-enhanced transformations of a polymer-bound base (PTBD) with phenols occur in less than 30 minutes (Scheme 78).[354]

Scheme 78

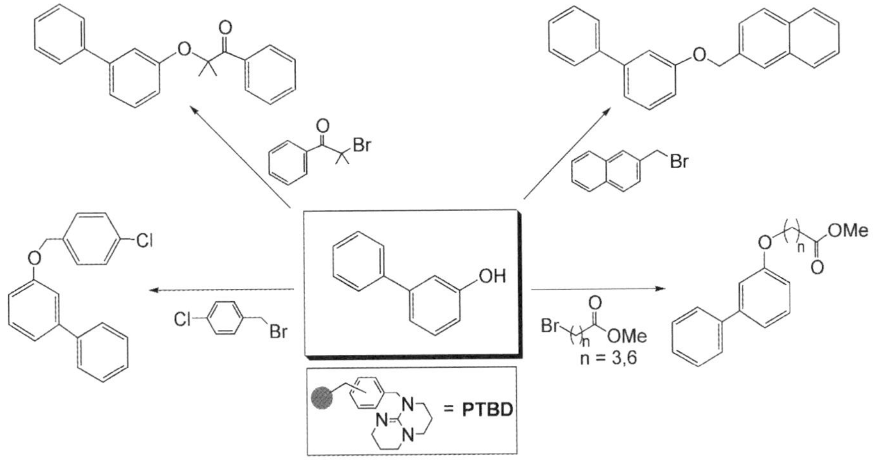

Microwave: 0.8 eq. alkyl halide, MeCN, **PTBD**, 175°C, 10-30 min. All reactions afforded 75-90%, based on LCMS.
Conventional: 25°C, 22-168 hours, 32-90% yield.

The Williamson etherification reaction is another O-alkylation of alcohols, both alkyl and aromatic. Synthetically, it is a simple method to both symmetrical and asymmetrical ethers, but it can also be used in the protection of alcohols. Normally, with thermal heat, the reaction of alcohol with a primary alkyl halide and a base catalyst can take up to twelve hours before completion. With microwave irradiation, etherification of *p*-cresol is accomplished in three minutes (Scheme 79) and of sesamol in four minutes (Scheme 80).[4,10,11,223]

Scheme 79

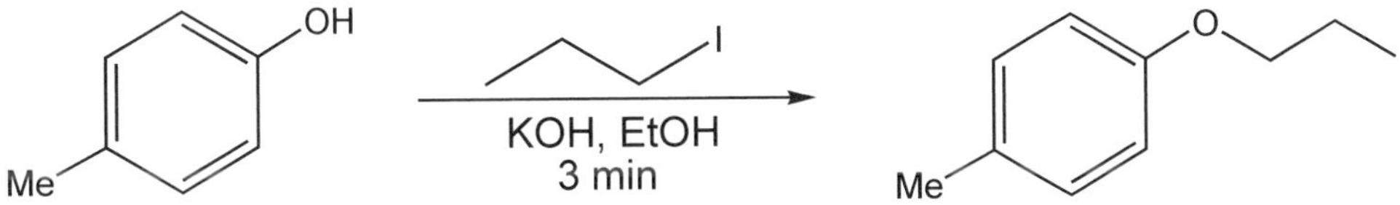

KOH, EtOH
3 min

Scheme 80

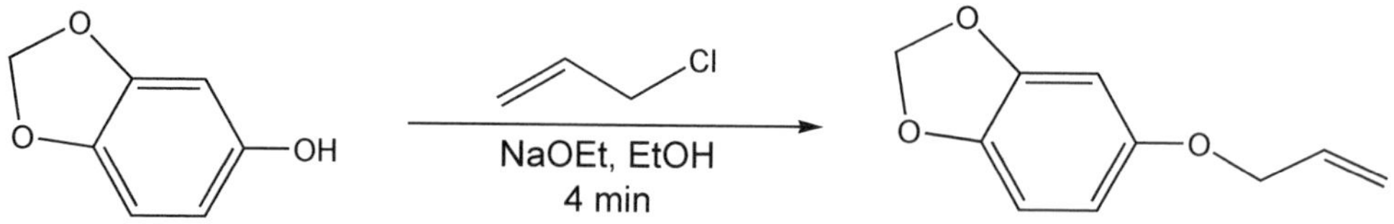

OH

Cl

NaOEt, EtOH
4 min

Esters are very important organic molecules in both the chemical and pharmaceutical industry. Esterification of carboxylic acids, alkylation of carboxylate anions, and transesterifications are the three types of methods for ester synthesis. Fisher esterification reactions are direct transformations of carboxylic acids in a sulfuric acid/alcohol mixture. With conventional heating, these conditions are harsh and can take anywhere from two hours to two days. Loupy et al., using microwave irradiation and *p*-toluenesulfonic acid (PTSA), provided esters in near quantitative yields in ten minutes or less (Scheme 81).[507]

Scheme 81

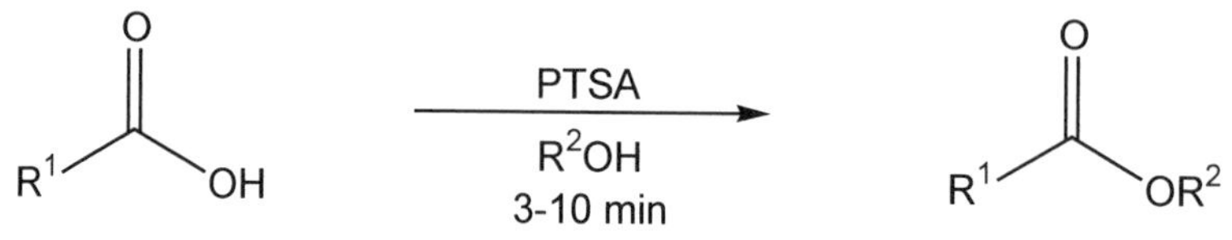

R^1–C(=O)–OH → R^1–C(=O)–OR^2

PTSA
R^2OH
3-10 min

R^1 = Ph, $(Me)_3Ph$, aliphatic chains
R^2 = Cy, Pr, Bu, and longer aliphatics

Alkylation of carboxylate anions is another routine transformation to esters. Once again, Loupy and colleagues have extensively examined this area of esterification.[505-507] Both potassium acetate (R_1 = Me) and potassium benzoate

(R_1 = Ph), first generated in situ with either potassium hydroxide or potassium carbonate, were mixed with different alkyl halides. Under thermal conditions, esters are achieved in five hours; however, microwave-driven substitutions proceeded in 5-15 minutes on alumina (Scheme 82). Loupy and co-workers have also investigated microwave irradiation in transesterifications reactions.[507] These reactions can be catalyzed by either an acid or a base, with PTSA and K_2CO_3, respectively, providing the most quantitative results. As shown in Scheme 83, the methoxy group of methyl benzoate is replaced by an octoxy group, which yields octyl benzoate and methanol. This reaction is successfully completed in two minutes with microwave heating.

Scheme 82

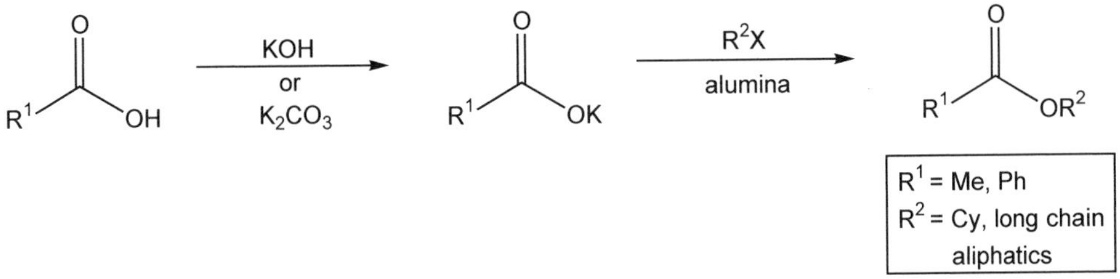
O
R1
OH
KOH
or
K2CO3
O
R1
OK
R2X
alumina
O
R1
OR2
R1 = Me, Ph
R2 = Cy, long chain
aliphatics

Scheme 83

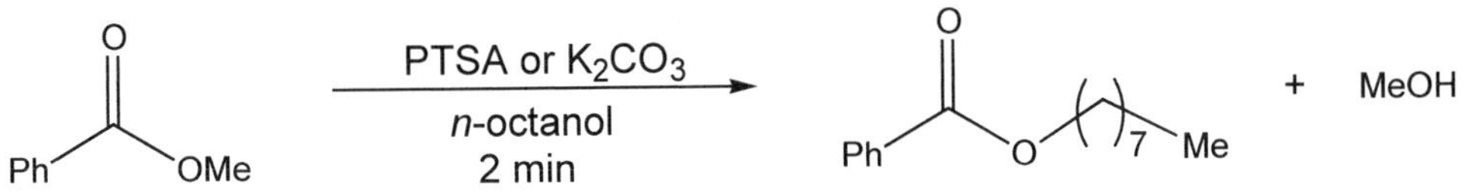
O
Ph
OMe
PTSA or K2CO3
n-octanol
2 min
O
Ph
O
7
Me
+
MeOH

The Finkelstein halogen exchange reaction is another nucleophilic substitution reaction. Alkyl iodides can be prepared easily from alkyl chlorides or bromides. This reaction is successful because, unlike sodium iodide, both sodium chloride and sodium bromide are not soluble in acetone or MEK. When an alkyl chloride or bromide is treated with sodium iodide, sodium chloride precipitates out of the solution, and formation of the alkyl iodide is favored. These reactions can take anywhere from 30 minutes to 80 hours for completion with conductive heating. Microwave heating yields alkyl iodides in ten minutes with excellent yields (Scheme 84).[10,223]

Scheme 84

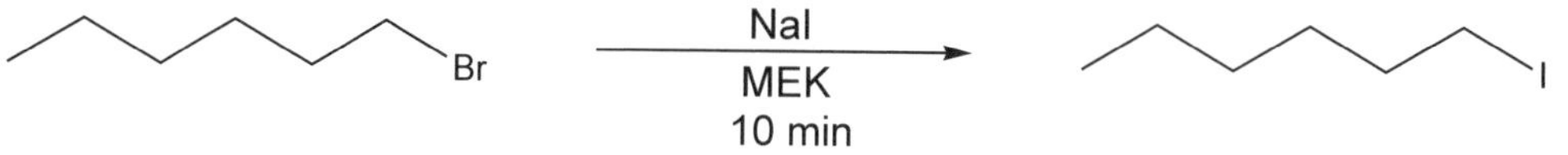

Br → (NaI; MEK; 10 min) → I

Another example of nucleophilic substitution reaction is radiolabeling. It is always a challenge to synthesize radiopharmaceuticals that are labeled with short-lived radionuclei. These reactions typically require long reaction times, and thus, have low radiochemical yields. The use of microwave irradiation provides shorter reaction times, and as a result, higher radiochemical yields.[519,520] S_NAr reactions, with ^{18}F-fluoride anion (via cyclotron), were performed on nitrobenzenes (Scheme 85).[519] Comparing conventional methods to that of microwave heating, the radiochemical yields, in most cases, more than doubled with the use of microwaves. Scheme 86 shows the synthesis of epi-[^{18}F]-fluoromisonidazole, which has been used in suspected cases of myocardial infarction.[520] Synthetically, the yields increased from 40% (conventional) to 65% overall with microwave irradiation. In addition, the entire route, including work-up, took less than 70 minutes with a 40% radiochemical yield.

Scheme 85

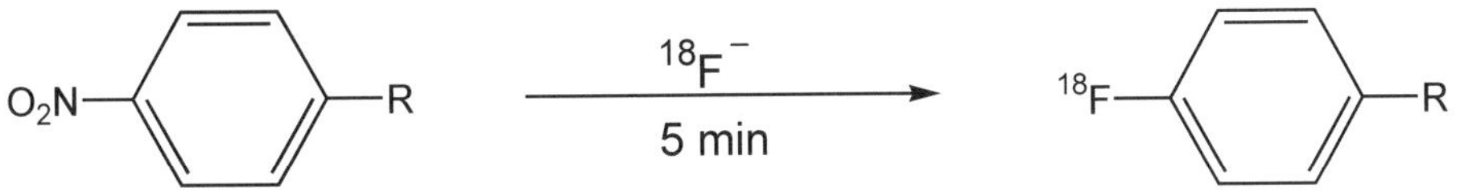

	Radiochemical yields		
	MW	Conventional	
R	5 min	5 min	30 min
CN	68	52	82
COMe	25	10	22
CO(cyclopropyl)	77	24	80

Scheme 86

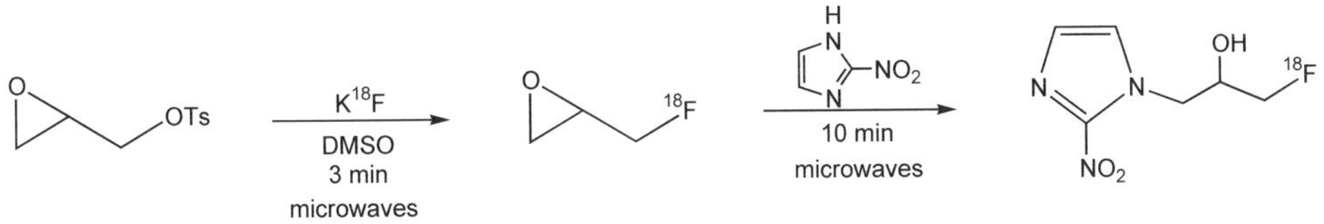

V. Electrophilic substitutions

Electrophilic substitutions also include a wide variety of different synthetic reactions. The majority of these reactions involve aromatic rings because most substitutions at an aliphatic carbon are nucleophilic. The converse is true with aromatic systems. They are more attracted to positively charged species because of their high electron density. Microwave irradiation has been used in examinations of Friedel-Crafts alkylations and acylations[223,521-525], sulfonylations[75,381,526,527], and deuterium-labeling[528] of aromatic rings.

Friedel-Crafts alkylations and acetylations are probably the most well known electrophilic aromatic substitutions. In these reactions, a proton directly attached to the aromatic ring is replaced with an alkyl or acetyl group. Lewis acids are necessary to promote the formation of a cationic intermediate, which the aromatic ring attacks. Reaction times vary with aromatic ring activity when using conductive heating. Electron-donating groups (EDGs) provide faster reactions, whereas rings with electron-withdrawing groups (EWGs) react much slower. Reactions that are heated with microwaves proceed in five minutes or less, regardless of substituents. Scheme 87 shows an unusual reaction between mesitylene and formaldehyde.[223] The carbocation that results is responsible for the Friedel-Crafts alkylation of another equivalent of mesitylene. This reaction proceeds in only four minutes with a 75% yield. Scheme 88 shows an intramolecular acylation on Bentonite ($Al_2O_3 \cdot 4SiO_2 \cdot H_2O$, montmorillonite) that yields anthraquinone in five minutes.[522]

Scheme 87

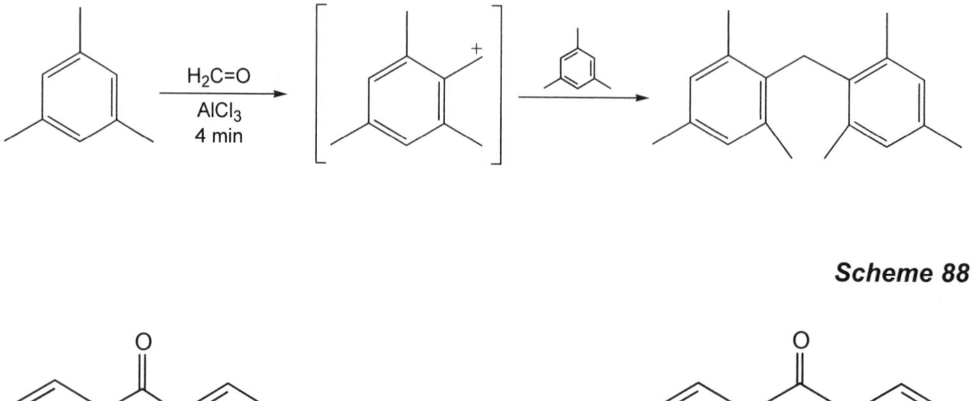

$H_2C{=}O$, $AlCl_3$, 4 min

Scheme 88

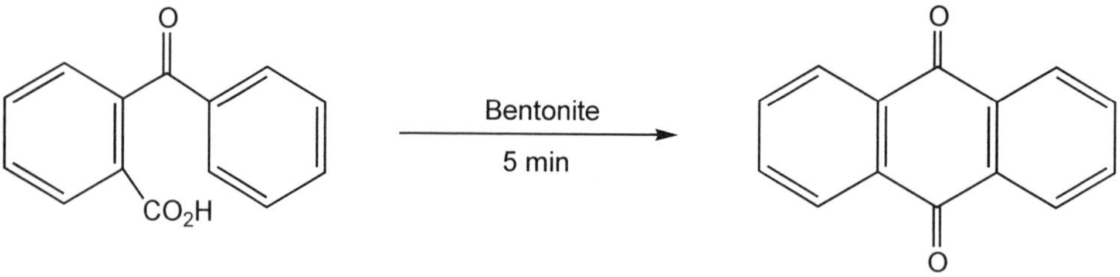

CO_2H — O — Bentonite, 5 min — O, O

Aromatic sulfonylation reactions are analogous to Friedel-Craft acylations. Generally, an aryl sulfonic acid chloride, coupled with a Lewis acid catalyst, reacts with an aromatic system to form a diaryl sulfone. The reaction can also be extended to the synthesis of alkyl aryl sulfones, from alkyl sulfonyl fluorides, and sulfonic acids, from sulfuric acid. With thermal conditions, these reactions usually require a stoichiometric amount of expensive catalyst and/or prolonged heating times. Marquie et al. have thoroughly examined electrophilic acylation and sulfonylation reactions under microwave irradiation.[523-526] They have synthesized diaryl sulfones, using a catalytic amount of inexpensive $FeCl_3$, in only five minutes with moderate to high product yields (Scheme 89).

Scheme 89

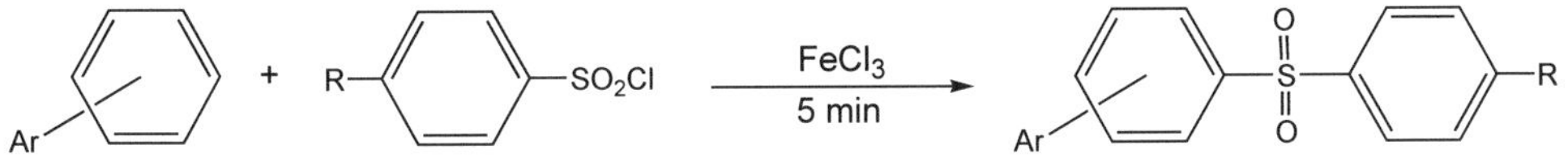

Ar + R–SO_2Cl → ($FeCl_3$, 5 min) Ar–SO_2–R

Ar = Me, *o*-$(Me)_2$, *m*-$(Me)_2$, *p*-$(Me)_2$
m-$(Me)_3$, Et, OMe, H, F, Cl, Br, I
R = H, Me, tBu, OMe, F, Cl

The sulfonylation of naphthalene is a very tricky transformation, regardless of reaction conditions. Both regioisomers, α- and β-substituted, are usually formed despite the fact that the β-substituted isomer is thermodynamically more stable. Interestingly, with temperature-controlled microwave heating for three minutes, α-naphthalenesulfonic acid is predominately formed in a 19:1 ratio (Scheme 90).[381]

Scheme 90

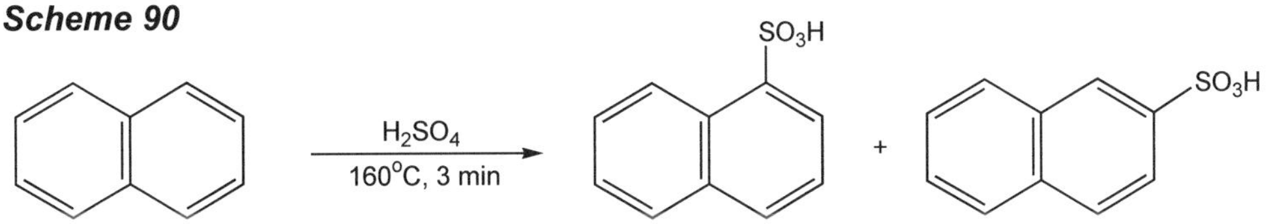

H_2SO_4, 160°C, 3 min

SO_3H + SO_3H

Ratio 19:1

Dehalogenation reactions are an effective and widely used method for deuterium-labeling (or tritium) of aromatic rings. Traditionally, labeling is achieved with either D_2 or T_2 gases. Both have solubility problems in organic solvents, and the storage of radioactive tritium waste is becoming an increasingly serious issue. Jones and co-workers have modified these labeling procedures by replacing D_2 and T_2 gases with labeled formates.[528] Scheme 91 exhibits a dehalogenation reaction on halogen substituted benzamide compounds. Utilizing deuterium labeled potassium formate, palladium(II) acetate, and either DMSO or ethanol as a solvent, coupled with microwave irradiation for 20 seconds, these reactions yielded 94% labeled product.

Scheme 91

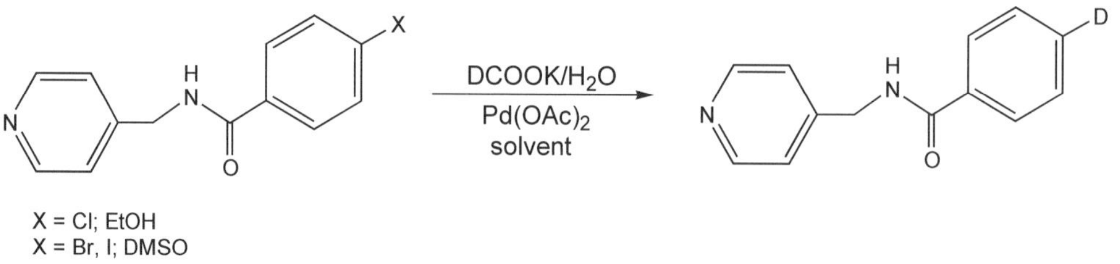

DCOOK/H_2O, $Pd(OAc)_2$, solvent

X = Cl; EtOH
X = Br, I; DMSO

I. Rearrangements

[3,3]-Sigmatropic rearrangements are important pericyclic reactions (concerted bond-making and -breaking). Two very important [3,3]-sigmatropic rearrangements are the Claisen and the Cope. In one type of Claisen rearrangement, an aryl vinyl ether rearranges to either an *ortho*-Claisen product or a *para*-Claisen product. In a Cope rearrangement, the products result from the rearrangement of a 1,5-hexadiene. Traditional methods for these transformations usually require very harsh reaction conditions, and in some cases, products will not form if the bulk temperature is

less than 200 °C. Utilization of microwaves decreases reaction times from days to minutes.[10,223,529,530] Scheme 92 shows an *ortho*-Claisen rearrangement in which both the reaction time and product yield were enhanced.

Scheme 92

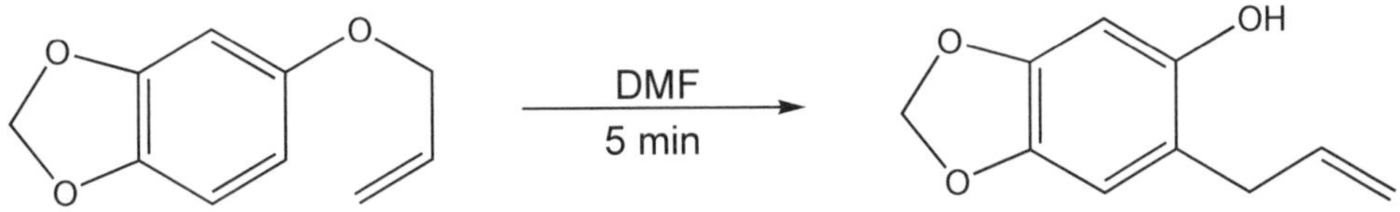

DMF
5 min

Microwave: 5 min, 97% yield
Conventional: 36 hr, 72% yield

In a *para*-Claisen rearrangement, an *ortho*-Claisen rearrangement is followed by a Cope rearrangement and tautomerization. In the example shown in Scheme 93, both the *ortho*-Claisen and the Cope are reversible

Scheme 93

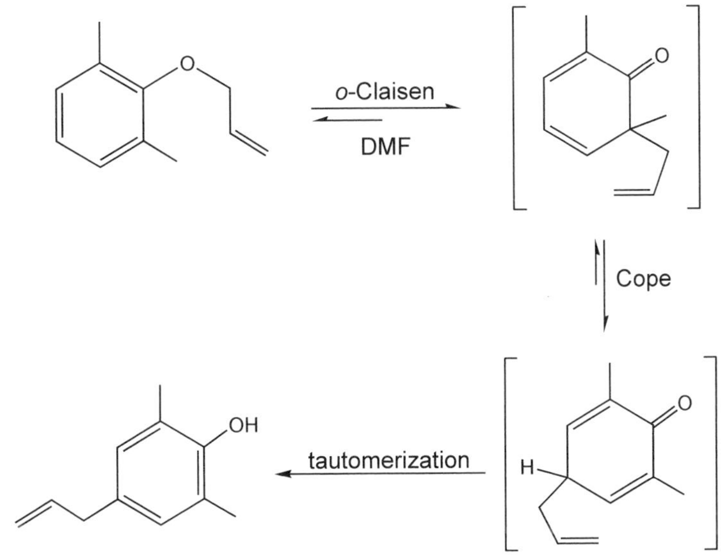

o-Claisen
DMF

Cope

tautomerization

Microwave: 20 min, 91% yield
Conventional: 4 days, 83% yield

transformations, but once tautomerization occurs to form the aromatic ring, the yield of the *para* product increases. Conventional methods usually require days of heating, but with microwave irradiation, the product is obtained in 20 minutes.[10,223]

The Fries rearrangement transforms an acyloxybenzene to an acylphenol. Acylphenols are important versatile organic intermediates that are used in agrochemical and pharmaceutical drug design. These reactions usually require stoichiometric amounts of Lewis acids and very long reflux times. In addition, they produce *ortho*/*para* mixtures. Moghaddam and co-workers have developed microwave-enhanced Fries rearrangements in dry media with 95% *ortho*-substituted products resulting (Scheme 94).[531] Incidentally, when cinnamyl esters of phenols were used, conjugate addition followed the rearrangement to yield flavanone derivatives (Scheme 95).[531]

Scheme 94

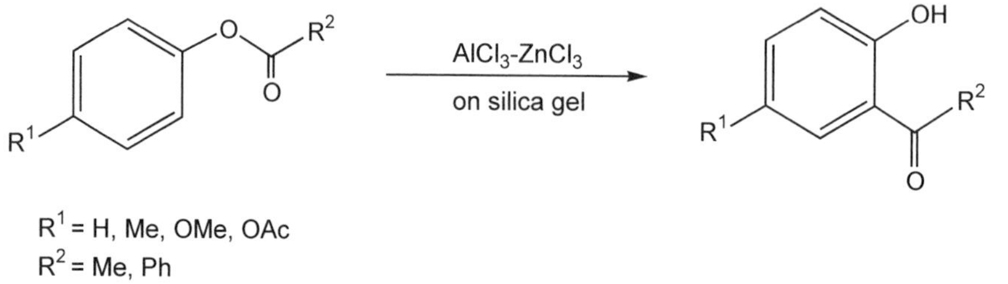

$AlCl_3$-$ZnCl_3$ on silica gel

R^1 = H, Me, OMe, OAc
R^2 = Me, Ph

Scheme 95

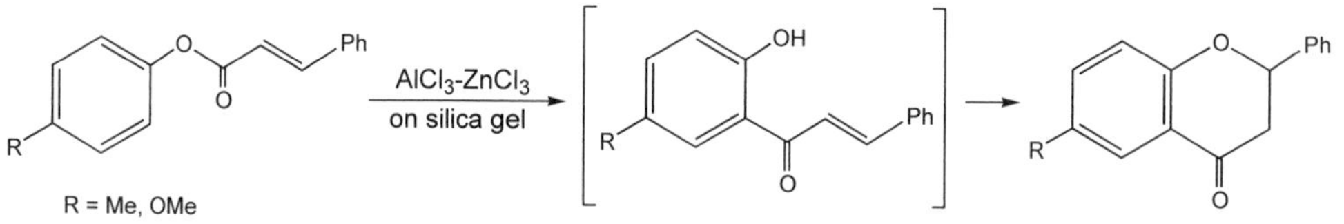

$AlCl_3$-$ZnCl_3$ on silica gel

R = Me, OMe

Moghaddam et al. have also developed a thia-Fries rearrangement where aryl sulfonates rearrange to phenolic sulfones (Scheme 96).[535]

Scheme 96

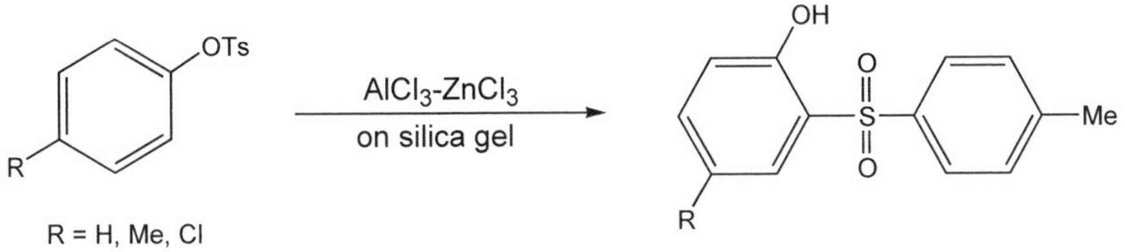

OTs

R

R = H, Me, Cl

$AlCl_3$-$ZnCl_3$

on silica gel

OH

O

S

O

Me

R

VII. Oxidations

Oxidation reactions are obviously quite important to synthetic organic chemistry. Traditionally, they require stoichiometric amounts of harsh, toxic oxidants and long reaction times. Microwave-induced oxidations have been extensively explored, including alcohols to carbonyl-containing compounds[137,160,536-567], as well as non-oxygen compounds like aromatics[568], sulfides[569], and carbon–carbon double bonds[570-572].

A very common reaction of alcohols is their oxidation to carbonyl compounds. Primary alcohols can yield aldehydes or carboxylic acids and secondary alcohols produce ketones. Tertiary alcohols generally will not yield any oxidized products. Numerous methods are available for oxidizing different types of alcohols. Scheme 97 shows microwave-induced oxidation of benzyl alcohols with various oxidants and methods.[10,223,543-551] In all of these reactions, microwave heating increased reaction rates drastically and also increased product yields.

Scheme 97

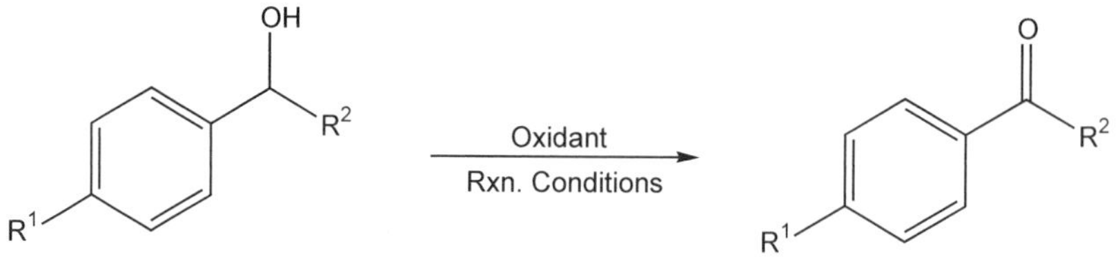

R^1 = H, *p*-Me, *p*-OMe, *p*-Et, *p*-NO_2
R^2 = H, Me, Et, Ph

Oxidant	Conditions	Microwave	Ref #
$PhI(OAc)_2$	alumina	1-3 min	547, 550
CrO_3	alumina	30 s - 5 min	544, 548
PCC	CH_2Cl_2	2 min	543
Clayfen	mont. clay	15 s - 1 min	551
MnO_2	silica gel	20 s - 1 min	549
MnO_2	Et_2O	7 min	9, 222

Benzoin oxidation to benzils, or 1,2-diketones, is another widely used reaction in synthetic chemistry. 1,2-Diketones are extremely important intermediates, as they can easily be transformed into many other organic functionalities. Conventional methods require extended reaction times with highly toxic oxidants. Using microwave irradiation eliminates both of these problems, as shown in Scheme 98.[564-567]

Scheme 98

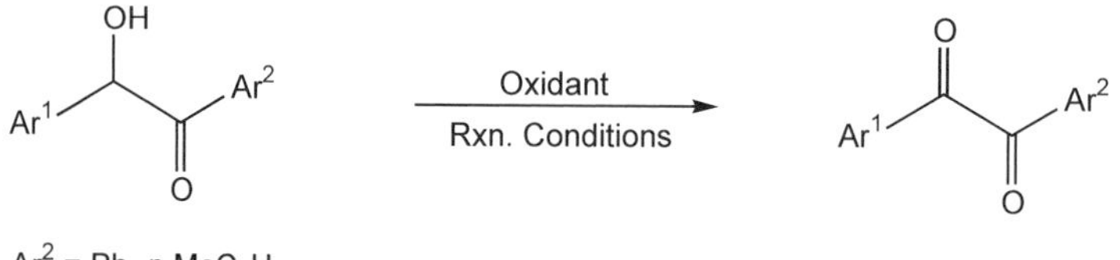

Ar^1, Ar^2 = Ph, *p*-MeC_6H_4, *p*-$MeOC_6H_4$, *p*-ClC_6H_4, furan

Oxidant	Conditions	Microwave/% Yield	Ref #
$CuSO_4$	alumina	2-4 min/81-96%	565
Oxone	alumina	2-3 min/71-88%	566
$Cu(OAc)_2/NH_4NO_2$	aq. AcOH	1-5 min/78-96%	567
PCC	$CHCl_3$	1 min/80-94%	567
HNO_3	neat	40 s - 1 min/79-95%	567

Aromatic tricyclic ring systems like fluorene, xanthene, diphenylmethane, and anthrone can also be oxidized. The methylene group in between the two aromatic rings can be directly oxidized to a carbonyl with potassium permanganate, but these reactions are frequently lengthy. With microwave irradiation, $KMnO_4$ on alumina, and solvent-free conditions, successful oxidation is completed in 10-30 minutes (Scheme 99).[568]

Scheme 99

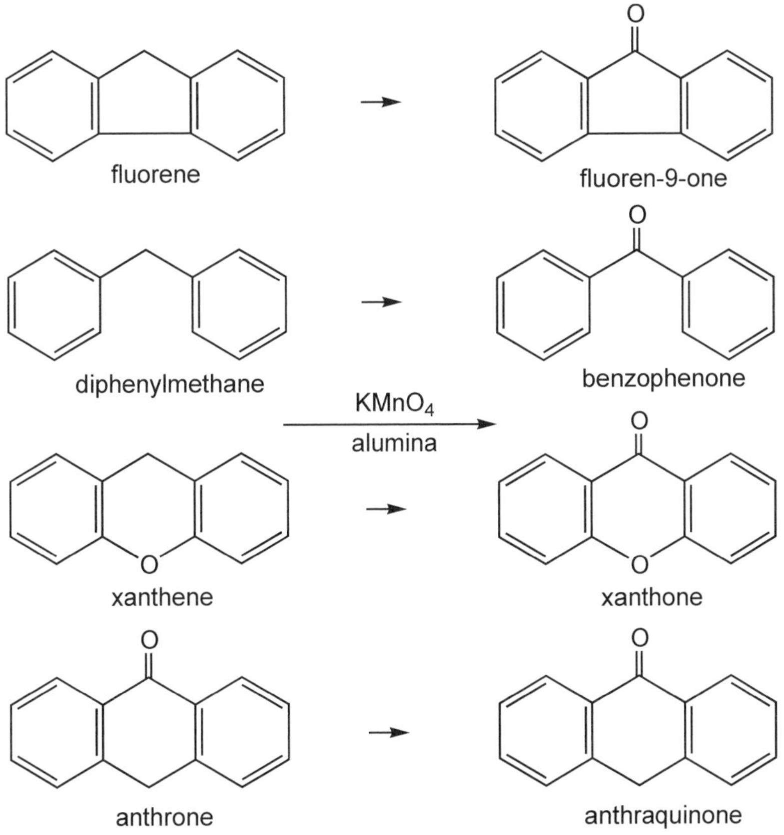

fluorene → fluoren-9-one

diphenylmethane → benzophenone

$KMnO_4$ / alumina

xanthene → xanthone

anthrone → anthraquinone

Sulfides can be readily oxidized to both sulfoxides and sulfones. To be selective for one or the other is the main challenge that faces organic chemists. Various methods have been employed, but most require long reaction times, the addition of extra reagent, and thus, a higher concentration of corrosive acids, peracids, or metallic compounds. Varma and co-workers, who have done extensive work with oxidants on supported mediums in solvent-free reaction environments[546-551,565,566,569], have performed selective oxidation on sulfides with sodium periodate on silica gel via microwave-enhanced reaction conditions (Scheme 100).[569]

Scheme 100

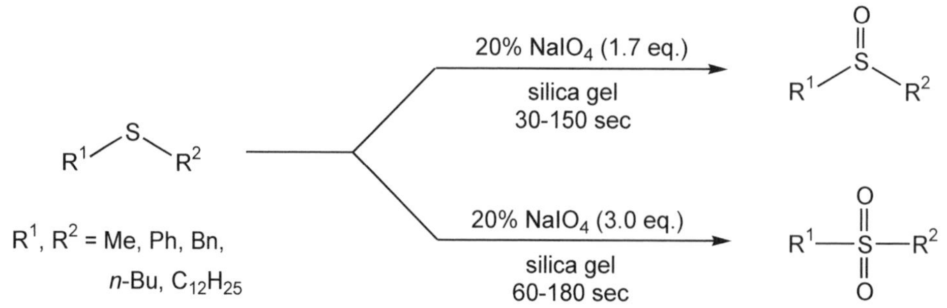

R^1, R^2 = Me, Ph, Bn, *n*-Bu, $C_{12}H_{25}$

20% $NaIO_4$ (1.7 eq.), silica gel, 30-150 sec

20% $NaIO_4$ (3.0 eq.), silica gel, 60-180 sec

Peracids are powerful oxidizing agents. Conventional use of *m*-chloroperbenzoic acid (MCPBA) on carbon–carbon double bonds forms epoxides, but these reactions are very long and usually take place at 0 °C. With microwave irradiation, epoxides are provided in only three minutes with 99% product yield (Scheme 101).[570]

Scheme 101

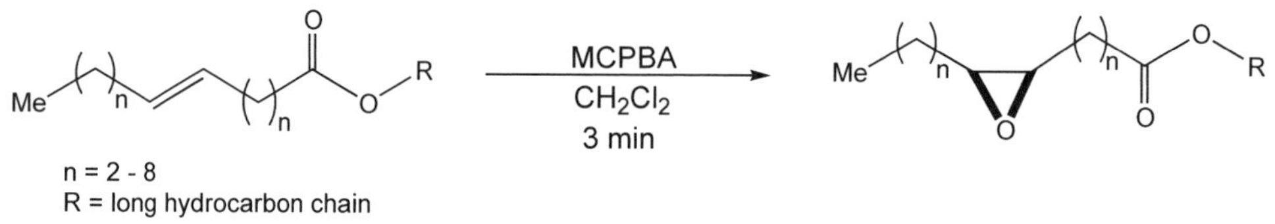

MCPBA, CH_2Cl_2, 3 min

n = 2 - 8

R = long hydrocarbon chain

VIII. Reductions

Reductions are very useful synthetic transformations and encompass a wide variety of applications. Microwave irradiation has been used to enhance yields and reaction rates of carbonyl to alcohol reductions[573-579], reductive amination of carbonyl compounds[580-582], aromatic nitro group reduction to amines[583], carbon–carbon double bond hydrogenations[584-587], and hydrogenolysis of functional groups[11,587-589].

Many versatile reagents can reduce ketones to secondary alcohols and aldehydes to primary alcohols. Sodium borohydride is widely used, as it is inexpensive, compatible with solvents, and safer to use than other reducing agents. Its major drawback is that solvents reduce its reaction rate and a large excess of the reagent is needed to successfully reduce any compound. Varma and co-workers have impregnated alumina with $NaBH_4$ and have reduced ketones and aldehydes with microwaves in a solvent-free environment (Scheme 102).[576] These reactions only required 1-5 equivalents of reducing agent and gave 80-93% product yields.

Scheme 102

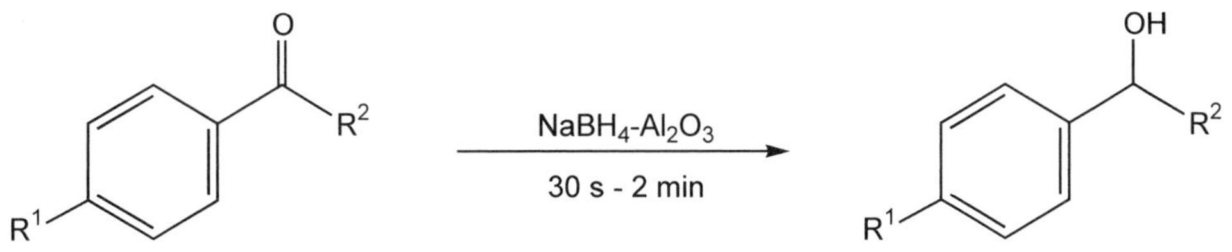

O
R²
$NaBH_4$-Al_2O_3
30 s - 2 min
OH
R²
R¹
R¹

R^1 = H, Me, OMe, Cl, NO_2
R^2 = H, Me, Ph

Reductive amination of carbonyl compounds is one of the most useful methods for synthesizing amines and their derivatives. The Borch reduction utilizes sodium borohydride derivatives for direct reduction to amines, while the Leuckart reaction produces N-formyl derivatives by using formamides. These reductions are plagued by high temperatures and long reaction times. Varma et al. used his $NaBH_4$ impregnated on support method, this time with clay, to effect reductive aminations on carbonyl compounds in five minutes or less.[582] Schiff bases (imines), generated in situ with microwaves on clay, followed by $NaBH_4$-clay addition, produce secondary amines in high yields (Scheme 103). Loupy and co-workers have synthesized N-formyl derivatives with microwave-enhanced Leuckart reactions in 30 minutes (Scheme 104).[581]

Scheme 103

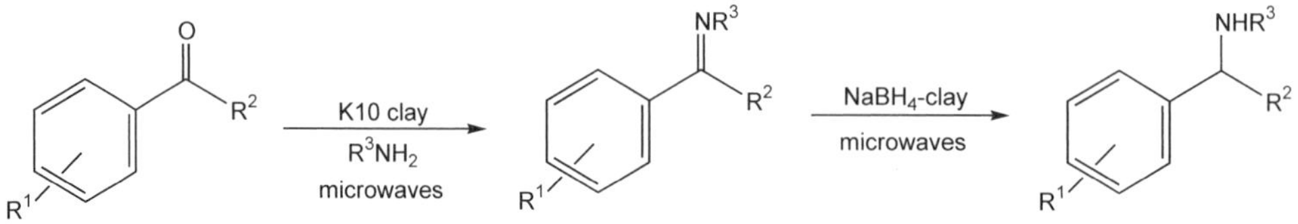

K10 clay, R^3NH_2, microwaves; $NaBH_4$-clay, microwaves

Scheme 104

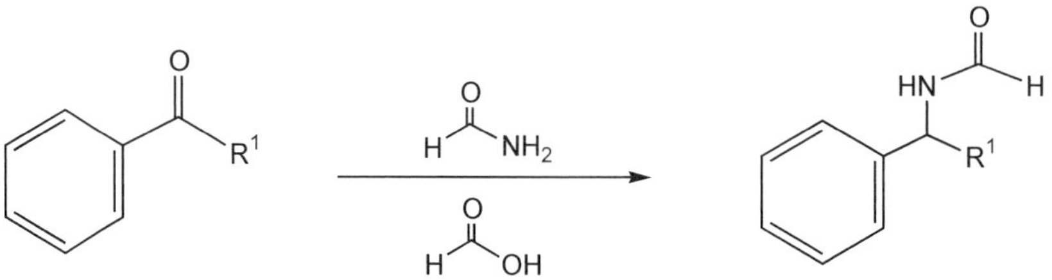

R^1 = Ph, Bn

R^1	% yield after 30 min MW	Conv.
Ph	99	2
Bn	99	12

Aromatic nitro groups can be reduced to amines by numerous methods in the solution phase. Conventional reaction conditions usually consist of the nitro compound, hydrazine hydrate, and a metal catalyst in refluxing ethanol or dioxane. Long reflux periods are required for successful reduction. As a solution, microwave-induced reduction with alumina-supported hydrazine and iron(III) chloride provided 100% conversion to aromatic amines (Scheme 105).[583]

Scheme 105

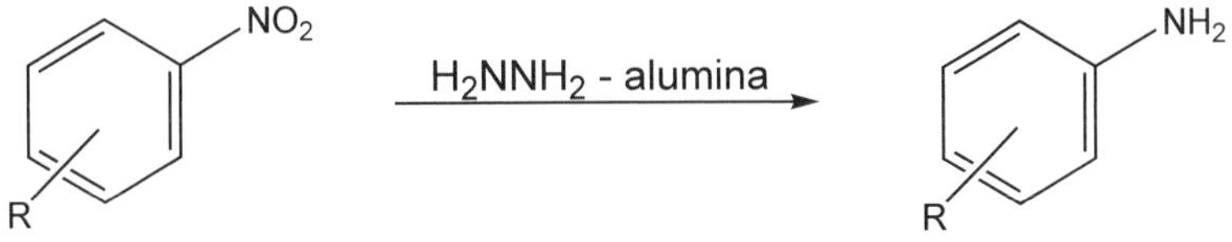

NO_2

H_2NNH_2 - alumina

NH_2

R

R

R = H, Me, OMe, OH, Cl, NH_2

In order to continue their research in microwave enhanced β-lactam synthesis, Bose and co-workers have developed hydrogenation and hydrogenolysis methods that utilize ammonium formate and either a Raney nickel or Pd/C catalyst (catalytic transfer hydrogenation).[587] As seen in Scheme 106, Raney nickel will only hydrogenate the carbon–carbon double bond, while use of a Pd/C catalyst will both hydrogenate and cleave the carbon–nitrogen bond (Scheme 107).

Scheme 106

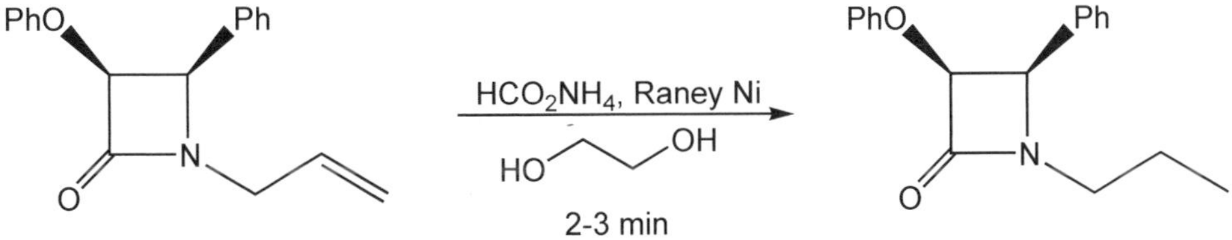

PhO Ph

O N

HCO_2NH_4, Raney Ni

HO OH

2-3 min

PhO Ph

O N

Scheme 107

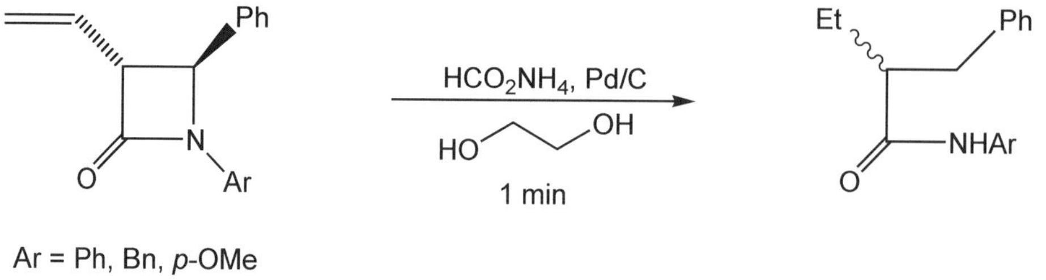

HCO_2NH_4, Pd/C; HO–CH2CH2–OH; 1 min

Ar = Ph, Bn, *p*-OMe

Complete reduction of a carbonyl group can be accomplished by the Wolff-Kishner reaction. Traditionally, these reactions require high temperatures and long reaction times. With microwave irradiation, reduction is completed in minutes with near quantitative product yields (Scheme 108).[588,589]

Scheme 108

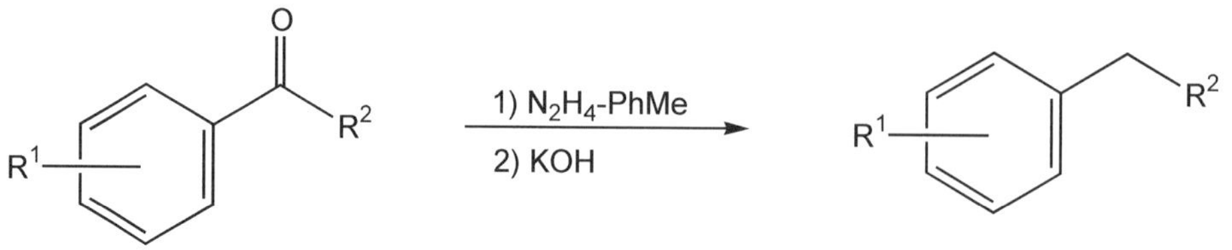

1) N_2H_4-PhMe
2) KOH

IX. Condensations

Condensation reactions are some of the most useful carbon–carbon bond forming methods available. However, many of these reactions require high temperatures and long reaction times. Microwave irradiation has been found to be quite effective in aldol[590-601], Knoevenagel[136,173,602-624], Pechmann[625], Henry[626,627], Mannich[628-630], and Ugi[631] condensations.

Knoevenagel reactions are generally base-catalyzed mixed aldol condensations. This reaction has been used successfully to synthesize coumarin derivatives.

These natural products are used extensively in fragrances, pharmaceuticals, and agrochemicals. Scheme 109 shows successful coumarin syntheses from microwave-enhanced reactions of hydroxy-aldehydes, esters, and a basic catalyst, piperidine.[605]

Scheme 109

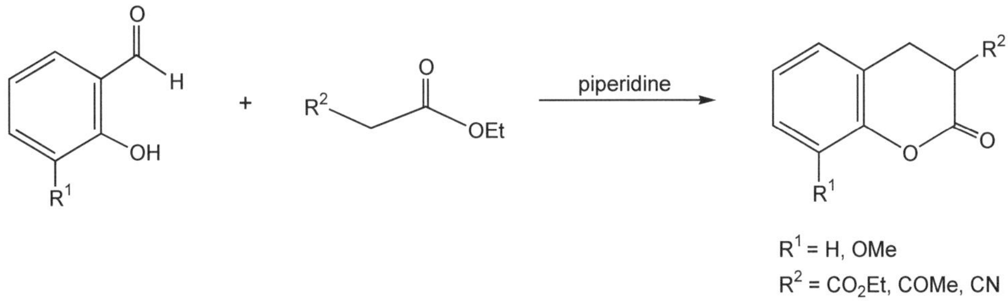

piperidine

R^1 = H, OMe
R^2 = CO_2Et, COMe, CN

Coumarins have also been synthesized by the Pechmann reaction, which involves a condensation of phenols with β-ketonic esters. Conventional Pechmann methods require harsh sulfuric acid conditions for a couple of days, and depending on reactivity of the substrates, high temperatures. Rare aminocoumarins can be synthesized on a graphite/montmorillonite K10 clay support in 5 to 30 minutes with microwave irradiation (Scheme 110).[625]

Scheme 110

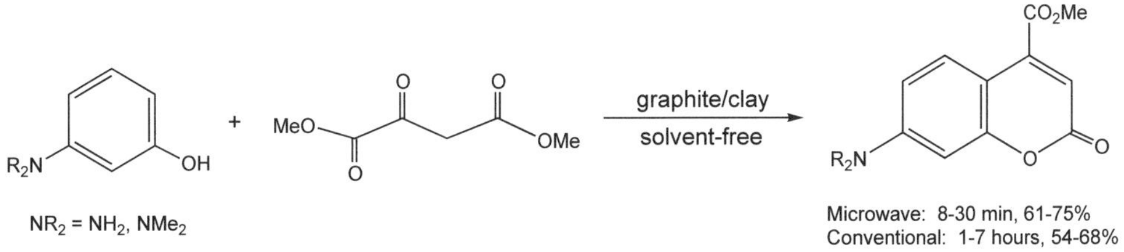

graphite/clay
solvent-free

NR_2 = NH_2, NMe_2

Microwave: 8-30 min, 61-75%
Conventional: 1-7 hours, 54-68%

Knoevenagel condensations can also be used in nitroalkene synthesis. Nitroalkenes are important synthetic building blocks to many potential pharmaceuticals. Scheme 111 exhibits a solid-phase approach to substituted nitroalkenes via microwave irradiation.[609]

Scheme 111

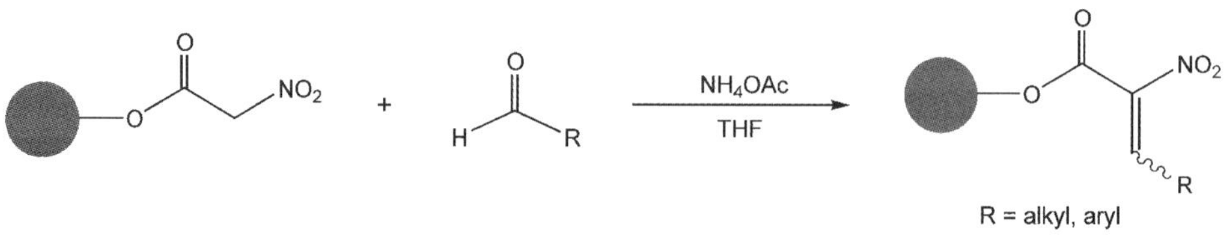

NO_2 + H R $\xrightarrow[THF]{NH_4OAc}$ NO_2 R

R = alkyl, aryl

Nitroalkenes can also be synthesized via the Henry condensation, which reacts a carbonyl compound with a nitroalkane under basic conditions. The resulting β-nitro alcohol dehydrates to give the nitroalkene. Classical conditions for the Henry reaction require elevated temperatures, which may not initiate the dehydration. Varma and coworkers have executed very high yielding solvent-free Henry reactions with a catalytic amount of ammonium acetate (Scheme 112).[626]

Scheme 112

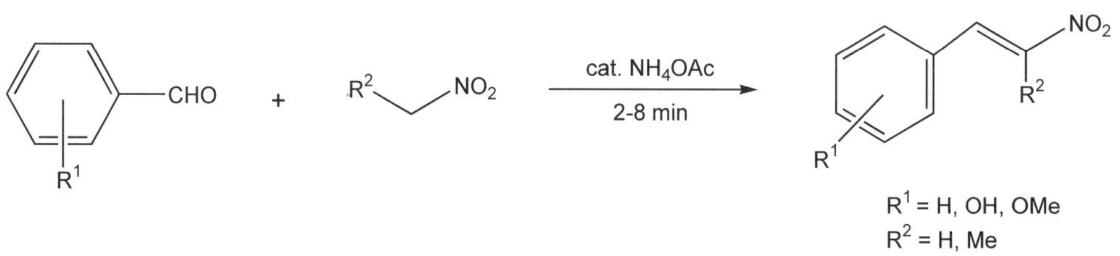

CHO + R^2 NO_2 $\xrightarrow[2\text{-}8\ min]{cat.\ NH_4OAc}$ NO_2 R^2 R^1

R^1 = H, OH, OMe
R^2 = H, Me

The Mannich condensation is generally a reaction between a carbonyl compound and an iminium ion, which is generated in situ from a secondary amine and formaldehyde. It is mainly used to introduce an α-dialkylaminomethyl substituent, and depending on the synthetic goal, it can then be thermally decomposed to α–methylene compounds. Traditional Mannich condensations often require severe reaction conditions. With microwave irradiation, Mannich reactions between *o*-ethynylphenols, secondary amines, and paraformaldehyde on CuI-doped alumina yield benzo[*b*]furans (Scheme 113).[629]

Scheme 113

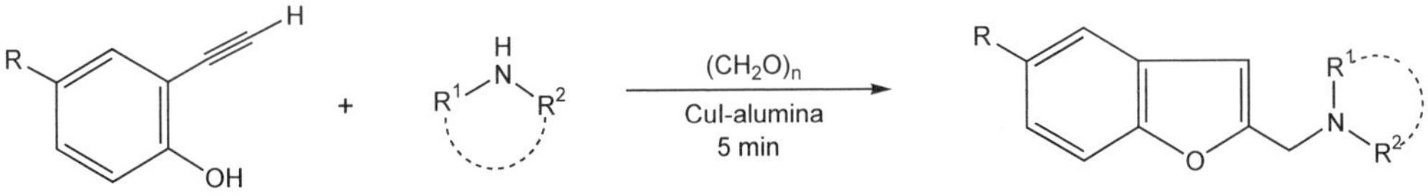

$(CH_2O)_n$, CuI-alumina, 5 min

R = H. Me. Ac

The Ugi reaction is another one-pot multi-component condensation reaction. There are four components used in this reaction, an amine, an aldehyde/ketone, a carboxylic acid, and an isocyanide. These combine to yield α-acylamino amides. Some Ugi reactions proceed rapidly, but most require 24 hours to several days for successful completion. Solid-phase Ugi condensations were effected in only five minutes in a 2:1 dichloromethane:methanol solvent mixture with microwaves (Scheme 114).[631]

Scheme 114

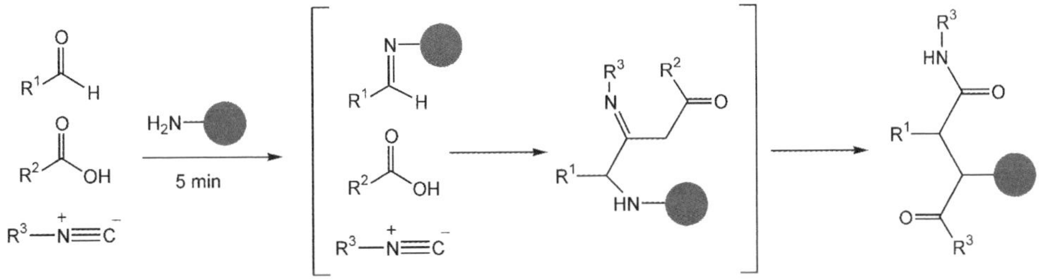

H_2N–, 5 min

X. Hydrolysis

Hydrolyses of organic compounds require the use of strong aqueous acids or bases and extensive periods of high conductive heating. Microwave irradiation is quite useful with this synthetic application in that it will hydrolyze esters, amides, nitriles, and peptides into carboxylic acids and their respective amines or alcohols in a very short period of time.[10,223,632-638] Scheme 115 shows both the acid- and base-catalyzed hydrolysis of acetanilide to aniline.[10,223] Ester hydrolysis (saponification) to a carboxylic acid is shown in Scheme 116.[638] Under basic conditions, 2-cyanotoluene can be hydrolyzed to both its carboxylic acid and amide derivative in a 5:95 ratio (Scheme 117).[10,223] Lastly, peptide hydrolysis, which normally can take twelve or more hours, is successfully completed in 15-30 minutes with microwave heating (Scheme 118).[10,223]

Scheme 115

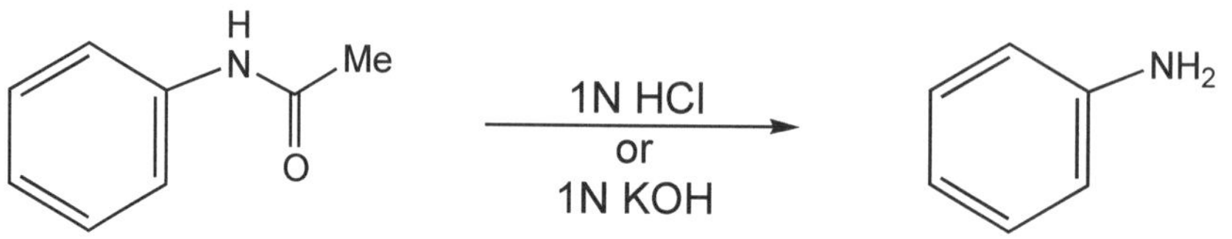

H N Me O — 1N HCl or 1N KOH → NH_2

	MW	Conv.
HCl	15 min, 91%	4 hrs, 98%
KOH	45 min, 83%	36 hrs, 60%

Scheme 116

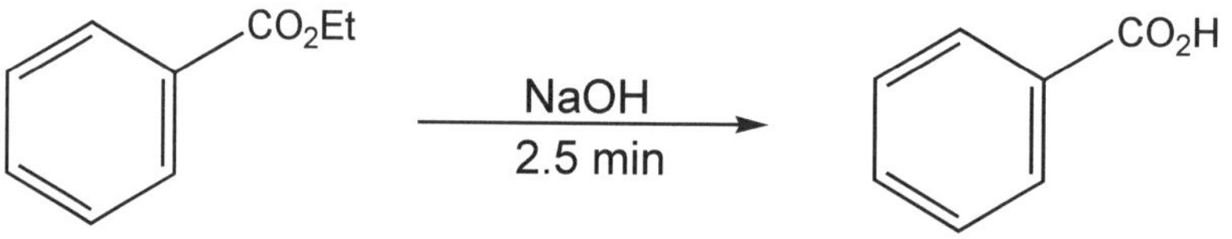

CO_2Et — NaOH, 2.5 min → CO_2H

Scheme 117

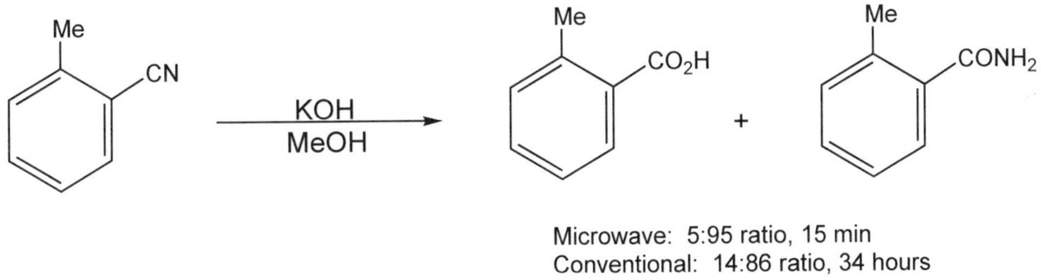

Scheme 118

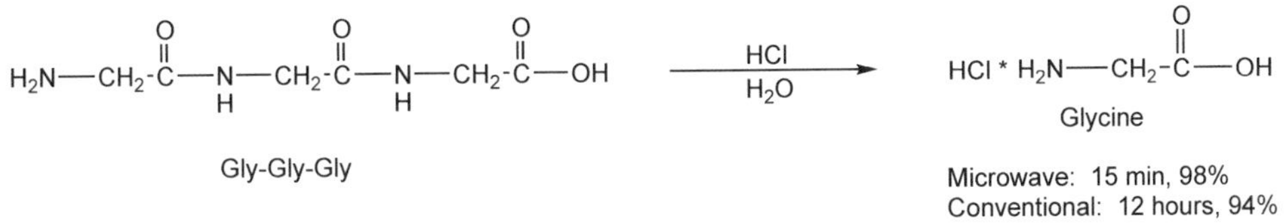

XI. Dehydration

Dehydration reactions also work exceptionally well with microwave irradiation. Simple dehydrations of alcohols to carbon–carbon double bonds[639-642], cyclodehydration reactions[643,644], and aldehydes to nitriles[93,645-651] can all be successfully executed in 1-15 minutes. Scheme 119 exhibits the dehydration of parthenin to anhydroparthenin.[639] Flavones can be synthesized from *o*-hydroxydibenzoylmethanes on clay by cyclodehydration (Scheme 120).[644]

Scheme 119

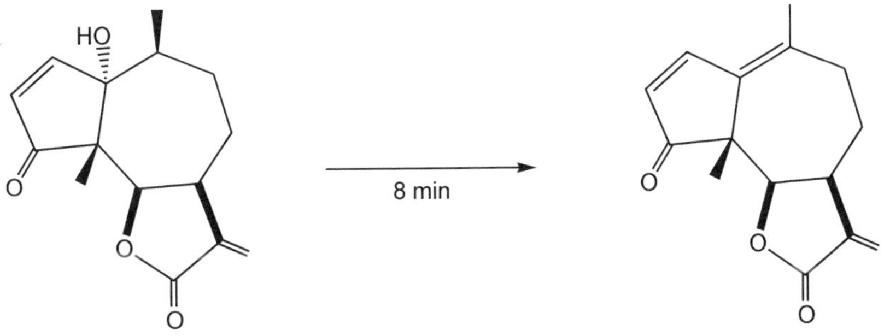

Aldehydes on bentonite clay can be converted to nitriles by using hydroxylamine hydrochloride and microwave irradiation (Scheme 121).[651]

Scheme 120

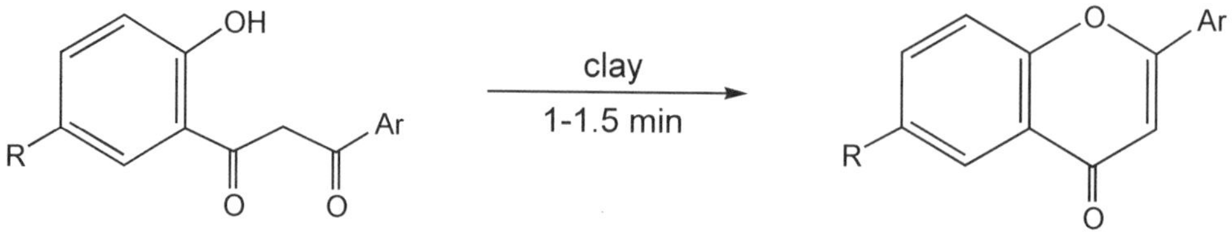

OH, Ar, R, O, O → clay, 1-1.5 min → O, Ar, R, O

Scheme 121

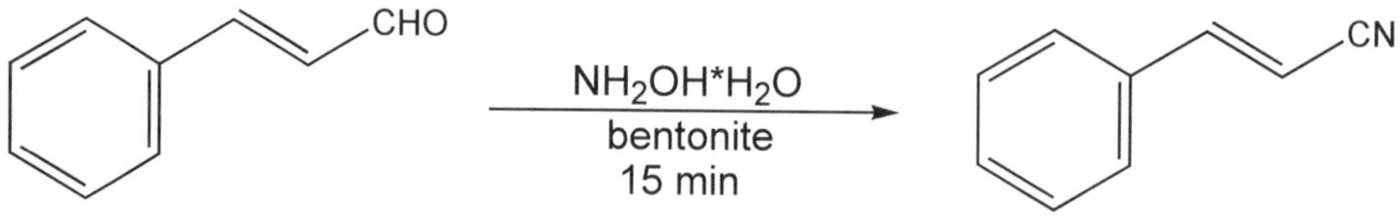

CHO → NH_2OH*H_2O, bentonite, 15 min → CN

XII. Protection and Deprotection

Two of the most important steps in synthetic organic chemistry routes are protection and deprotection of important functional groups. Protecting groups are needed to temporarily block a certain reactive site on a molecule. The protective group is then chemically removed (deprotected) in a later step and that particular reactive functional group is regenerated. There are many different methods of both protection and deprotection. Traditionally, protection has selectivity problems, while harsh conditions are needed for deprotection. Microwave irradiation has been shown to be quite effective in both of these areas, as well as decreasing reaction times immensely. Trifluoroacetylation of amines and esterification of alcohols were both briefly discussed in Section IV of this chapter. In addition, research has been performed on the protection of amino acids[653], etherification of alcohols[654-656] and diols[657], and both oxathiolane/dithiolane[62,150,658,659] and ketal/acetal[62,150,660-667] protection of carbonyl compounds.

Hydroxyl groups are present in a large number of compounds that are of pharmaceutical interest, including nucleosides, carbohydrates, steroids, and macrolides. They are susceptible to oxidation, acetylation, and halogenation, and therefore, must be protected occasionally in synthetic routes. Etherification is one of the most widely used alcohol protection methods available. Traditionally, the selective protection of one of two identical hydroxyls in a symmetrical molecule is quite limited. Successful monotetrahydropyranylation has been effected on symmetrical diols in less than three minutes with microwave irradiation (Scheme 122).[654] Acetals and ketals are used to protect 1,2- and 1,3-diols. Scheme 123 shows acetalization effected on clay in a solvent-free environment with ten minutes of microwave heating.[657]

Scheme 122

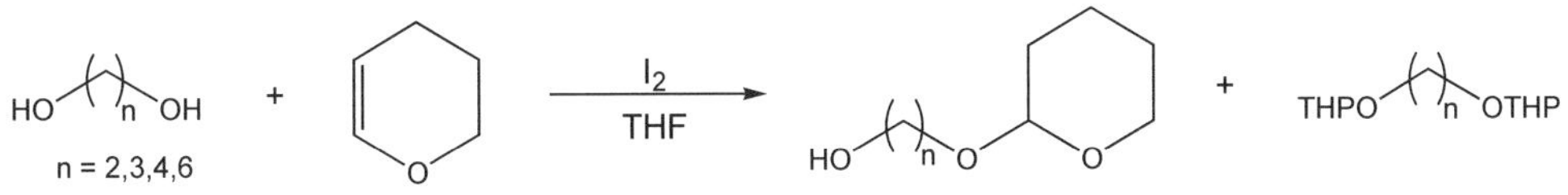

HO–(CH₂)n–OH + dihydropyran → (I_2, THF) → HO–(CH₂)n–O–THP + THPO–(CH₂)n–OTHP

n = 2,3,4,6

	mono	diether
MW	75-78%	15-17%
Conv.	43%	51%

Scheme 123

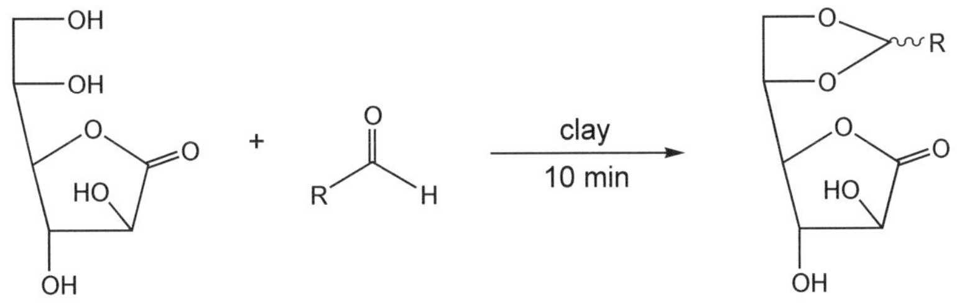

clay

10 min

Microwave: 60-66%
Conventional: 22-38%

Carbonyls are very susceptible to nucleophilic attack. Acyclic and cyclic acetals or ketals are the most widely used protection method for carbonyl-containing compounds. Classically, this reaction is usually acid catalyzed and requires high temperatures for the azeotropic removal of water with a Dean-Stark trap. Successful acetalization resulted on aldehydes and ketones by using a catalytic amount of *p*-toluenesulfonic acid coupled with microwave irradiation (Scheme 124).[660]

Scheme 124

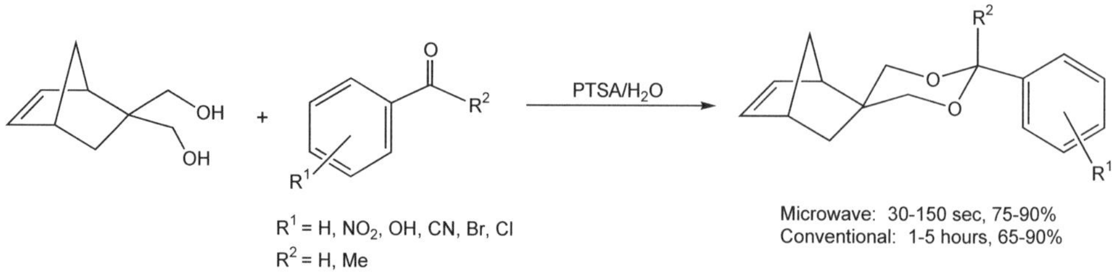

OH OH + R^1 R^2 O — PTSA/H_2O → R^2 O O R^1

R^1 = H, NO_2, OH, CN, Br, Cl
R^2 = H, Me

Microwave: 30-150 sec, 75-90%
Conventional: 1-5 hours, 65-90%

Extensive work has been done on deprotection methods that have required harsh conventional reaction conditions. Successful research on deprotection methods to alcohols or diols include demethylation[668], deacetylation[669,670], and oxidative deprotection of ethers[83,671-682]. Cleavage of oxathiolanes[683], thioacetals[684,685], acetals[82,123,686-689], hydrazones[690-692], semicarbazones[692-694], and oximes[69,76,79,80,146,170,695-700] with microwave irradiation provides carbonyl-containing compounds readily. Lastly, deprotection of esters yields carboxylic acids.[701-706]

To reiterate, ethers are the most widely used protective group in synthetic chemistry. Their high stability against a variety of reaction conditions makes them very effective. Deprotection of methylated phenolic derivatives is hard to accomplish mildly, and therefore, requires acidic reagents and high temperatures. Use of pyridine hydrochloride and microwave irradiation successfully regenerates phenols in 15 minutes (Scheme 125).[668]

Varma et al. show how both deacetylation[669] (Scheme 126) and cleavage of *t*-butyldimethylsilyl ethers[679] (Scheme 127) can easily be achieved directly on alumina in solvent-free reaction conditions. Traditionally, deprotection of THP ethers to their respective alcohols is achieved with toxic chromium(VI) reagents. Heravi and co-workers have used iron(III) nitrate on clay for direct oxidation of THP ethers to their carbonyl compounds (Scheme 128).[676]

Scheme 125

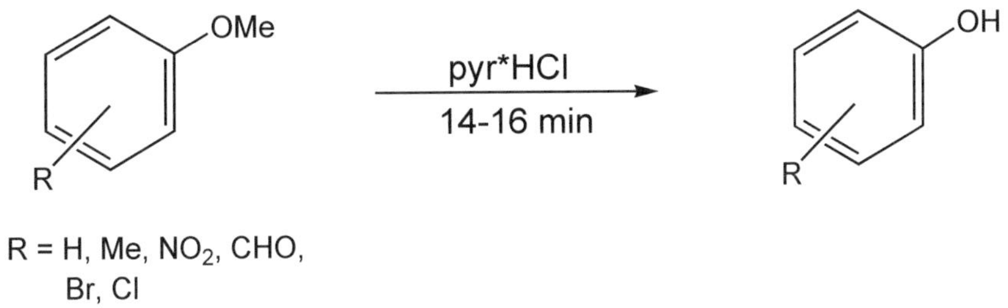

OMe
pyr*HCl
14-16 min
OH
R
R

R = H, Me, NO_2, CHO, Br, Cl

Scheme 126

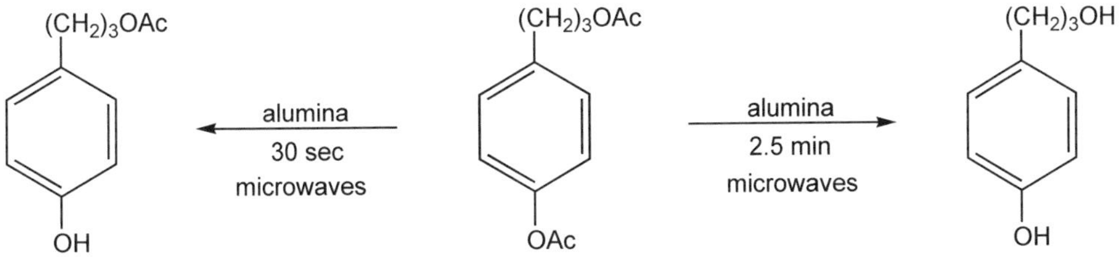

$(CH_2)_3OAc$
OH
alumina
30 sec
microwaves
$(CH_2)_3OAc$
OAc
alumina
2.5 min
microwaves
$(CH_2)_3OH$
OH

Scheme 127

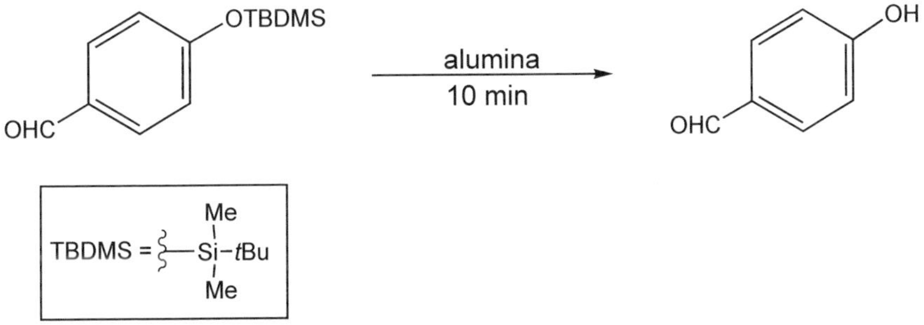

OTBDMS
OHC
alumina
10 min
OH
OHC

TBDMS = Si(Me)(Me)-*t*Bu

Scheme 128

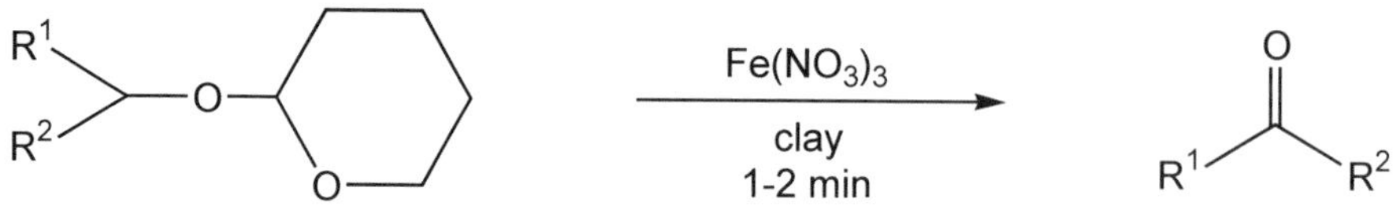
R1
R2
O
O
Fe(NO3)3
clay
1-2 min
O
R1
R2

Dithioacetals and acetals, both cyclic and acyclic, are superb carbonyl protecting groups and are used extensively in synthetic routes. The sulfur-containing acetals are very effective, as they are highly stable against strong acids and bases. Conventional deprotection methods usually require toxic heavy metals. Successful dethioacetalization of thioacetals/ketals, utilizing iron(III) nitrate on clay (clayfen) and very little microwave irradiation, occurred with 87-98% product yields (Scheme 129).[684] Diacetyl acetals are very efficient as protectors of the aldehyde moiety. Geminal diacetates are quite stable in acidic conditions and are cleaved by strong bases. Traditionally, deprotection is executed by either overnight stirring with sodium hydroxide or refluxing in alcoholic sulfuric acid. Scheme 130 exhibits deacetalization of benzaldehyde diacetate derivatives on neutral alumina in 30-40 seconds, 88-98% yields.[689]

Scheme 129

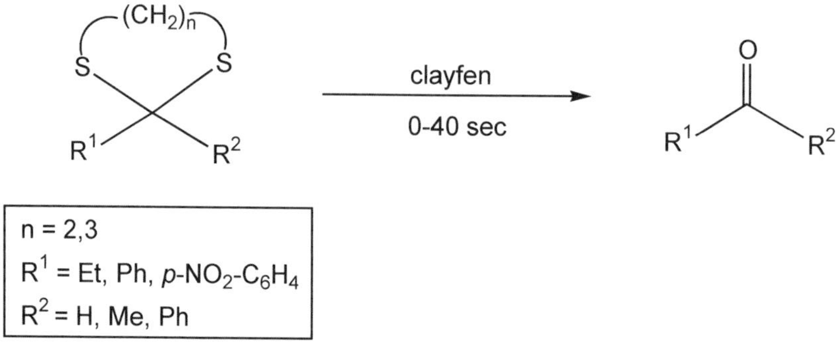
(CH2)n
S
S
R1
R2
clayfen
0-40 sec
O
R1
R2
n = 2,3
R1 = Et, Ph, p-NO2-C6H4
R2 = H, Me, Ph

Scheme 130

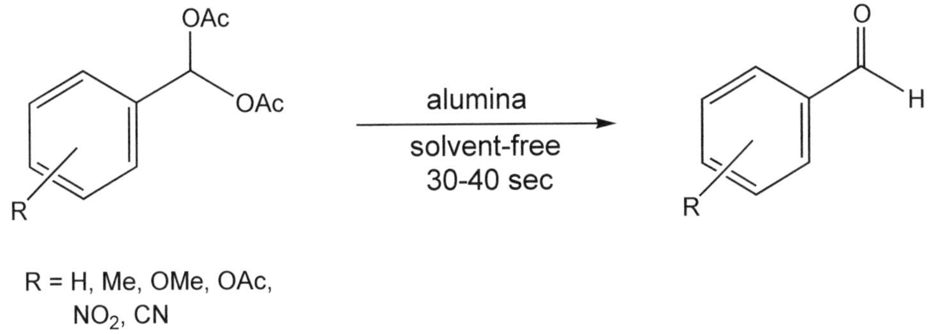

OAc
OAc
alumina
solvent-free
30-40 sec
O
H
R
R

R = H, Me, OMe, OAc, NO_2, CN

Semicarbazones, hydrazones, and oximes are essentially functional group equivalents of carbonyl compounds, and thus, are useful protecting groups. Conventional deprotection of these derivatives usually requires long reaction times with very high temperatures. Semicarbazones and hydrazones can be deprotected to regenerate the carbonyl with ammonium persulfate on clay coupled with microwave irradiation (Scheme 131).[692] Microwave-enhanced regeneration of the carbonyl by deoximation can be achieved by use of either ammonium persulfate[698] or sodium periodate[699] on silica, or even pyridinium chlorochromate[700] (Scheme 132).

Scheme 131

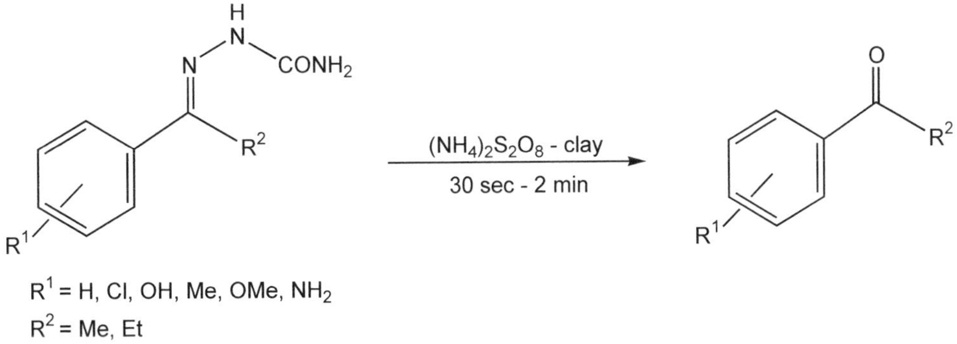

H
N
N
$CONH_2$
R^2
$(NH_4)_2S_2O_8$ - clay
30 sec - 2 min
O
R^2
R^1
R^1

R^1 = H, Cl, OH, Me, OMe, NH_2
R^2 = Me, Et

Scheme 132

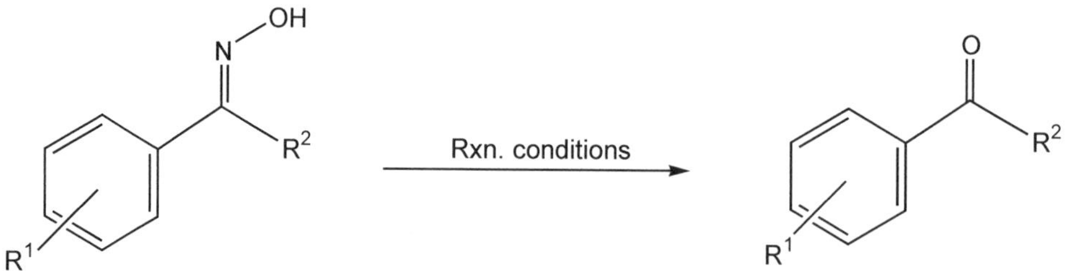

Rxn. conditions

Method	Microwave	Yield
$(NH_4)_2S_2O_8$ - silica	1-2.5 min	60-83%
$NaIO_4$ - silica	1-2 min	68-93%
PCC/CH_2Cl_2	2 min	90-97%

Carboxylic acids mainly need to be protected in order to mask the acidic proton in base-catalyzed reactions. Esters are a useful protecting group for the carboxyl moiety, as they remove the acidic proton and provide for easier handling of the molecule. Deesterification traditionally provides moderate yields and poor chemoselectivity. Varma and co-workers have successfully deprotected benzyl esters in solvent-free conditions on alumina (Schemes 133 and 134).[702]

Scheme 133

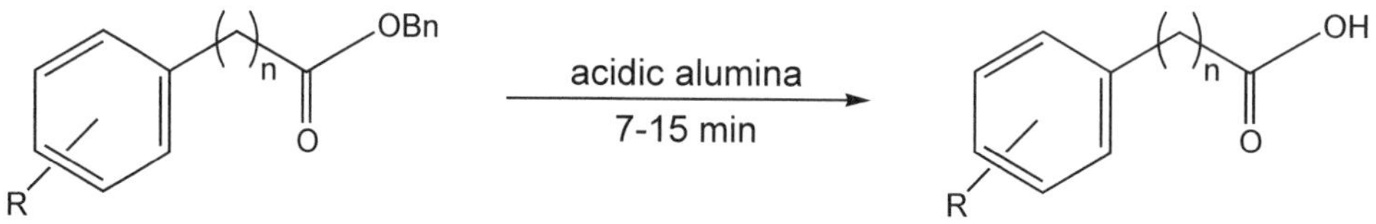

acidic alumina
7-15 min

n = 0,1,2
R = H, OH, OMe, OBn

Scheme 134

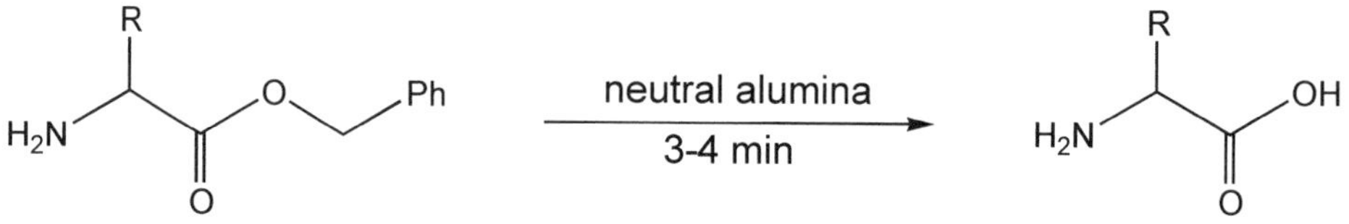

neutral alumina
3-4 min

XIII. Miscellaneous reactions

Isomerizations of double bonds and tautomerizations (interconversion of isomers) are useful organic transformations. These also can be enhanced by microwave irradiation. Loupy et al. have used microwave-induced, solvent-free, solid-liquid-phase transfer catalysis(PTC), which employs a salt (in this case, potassium *t*-butoxide) and a phase transfer catalyst (tetrabutylammonium-bromide, TBAB), to isomerize eugenol to isoeugenol (Scheme 135).[708] An example of a microwave-enhanced double bond isomerization followed by enol – keto tautomerization is shown in Scheme 136.[10,223]

Scheme 135

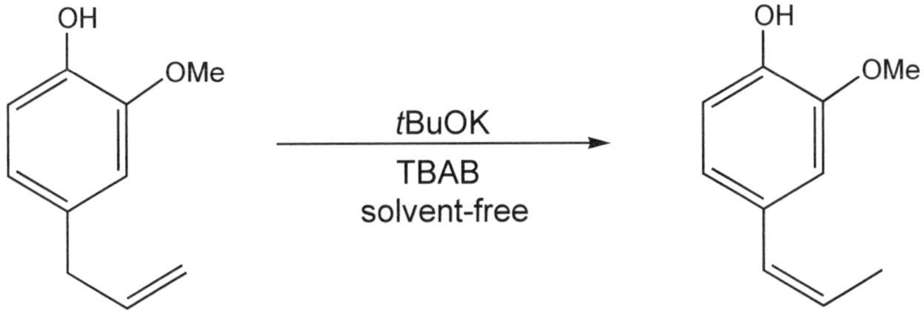

OH OMe — *t*BuOK / TBAB / solvent-free → OH OMe

Scheme 136

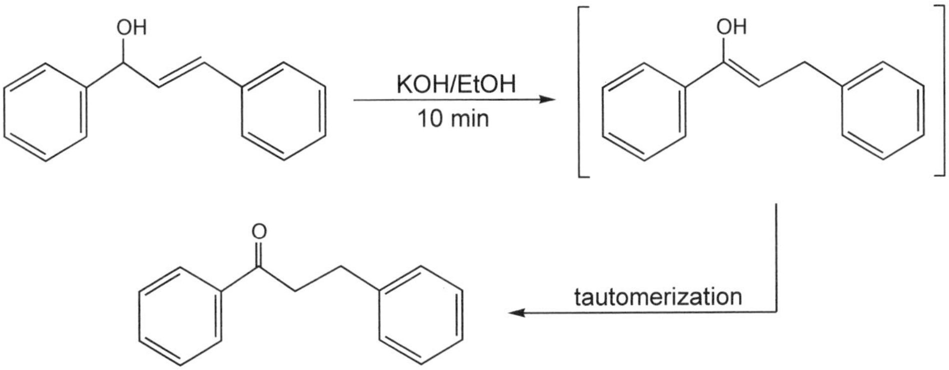

OH — KOH/EtOH / 10 min → [OH] — tautomerization → O

Microwave: 10 min, 97% yield
Conventional: 20 hrs, 86% yield

Dealkoxycarbonylation, also known as the Krapcho reaction, completely removes an ester group directly from the carbon alpha to a carbonyl. These reactions are difficult to achieve with conductive heating, and they usually require very high temperatures with DMSO as the solvent. Loupy uses PTC with LiBr/TBAB to transform malonic esters to monoesters (Scheme 137).[709]

Scheme 137

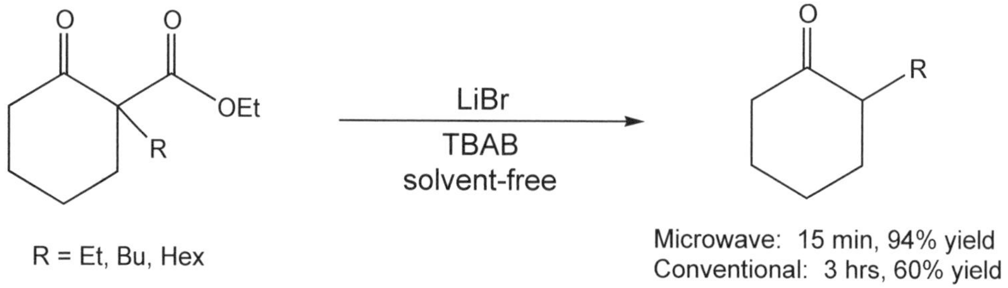

O O OEt R

LiBr

TBAB

solvent-free

O R

R = Et, Bu, Hex

Microwave: 15 min, 94% yield
Conventional: 3 hrs, 60% yield

Hydrosilylation of alkenes is another reaction that proceeds poorly with conventional heating. The example shown below in Scheme 138 normally requires 18 hours of thermal heat and provides only a 5% product yield. With six 30-second bursts of microwave irradiation, the 2-vinylpyridine is silylated with a 75% yield of product.[381]

Scheme 138

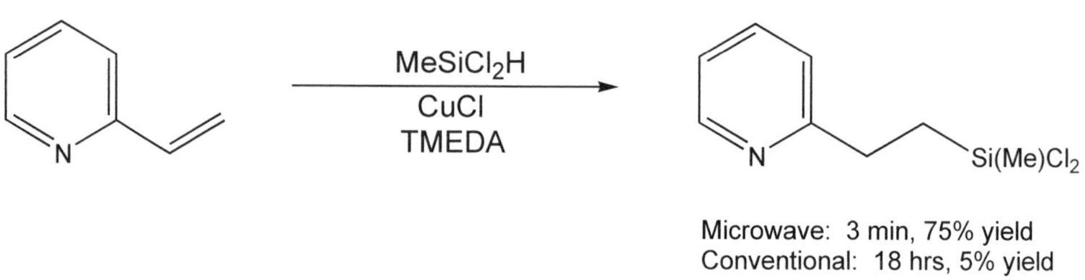

N

$MeSiCl_2H$

CuCl

TMEDA

N $Si(Me)Cl_2$

Microwave: 3 min, 75% yield
Conventional: 18 hrs, 5% yield

The Wittig reaction is probably the most reliable olefin-forming reaction in synthetic organic chemistry. In this reaction, an aldehyde or a ketone reacts with a phosphorus ylide, forming an oxaphosphetane intermediate. The four-membered ring collapses and produces the alkene and a phosphine oxide byproduct. Wittig reactions, traditionally, can require up to 24 hours of reflux in high boiling point solvents. There has been some research performed on the microwave-assisted synthesis of both the Wittig reagent (ylide)[710] and the reaction itself [711-719]. Scheme 139 exhibits successful Wittig transformations on benzaldehyde derivatives that occurred in five minutes.[711]

Scheme 139

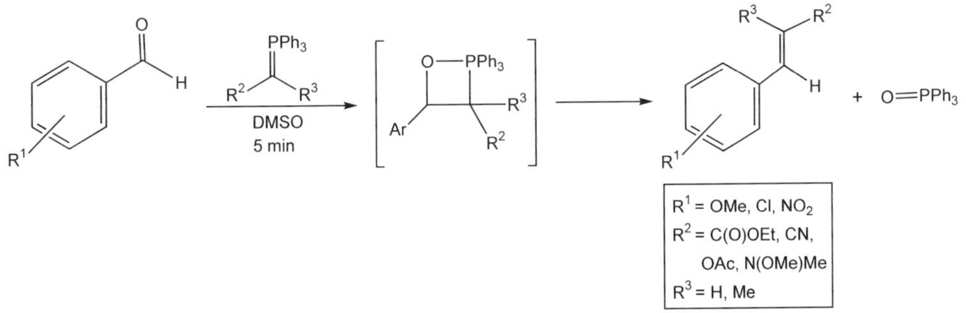
PPh3
DMSO
5 min
O—PPh3
Ar
+ O=PPh3
R1 = OMe, Cl, NO2
R2 = C(O)OEt, CN, OAc, N(OMe)Me
R3 = H, Me

The Peterson olefination, also known as the silyl-Wittig reaction, utilizes a trialkylsilylmethyl lithium (or magnesium) reagent that adds to a ketone or aldehyde. The β-hydroxysilane intermediate, which can be isolated, eliminates water upon acid or base catalysis. Depending on the stability of the β-hydroxysilane, these reactions can require many hours of reflux. A Peterson reaction in which the silylmethyl anion was generated in situ with cesium fluoride on clay was successful in ten minutes with microwave heating (Scheme 140).[721]

Scheme 140

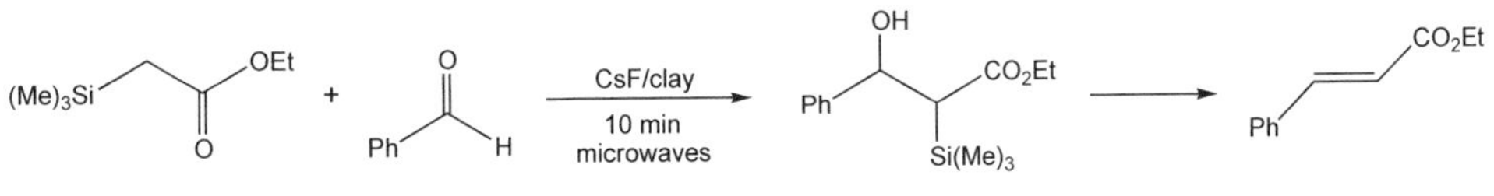
(Me)3Si
OEt
Ph
H
CsF/clay
10 min
microwaves
OH
CO2Et
Si(Me)3
CO2Et

The role of microwave synthesis in drug discovery and development will only increase over the next several years. There is a need for a very simple, flexible, and compact microwave system that can be used in synthesis laboratories. As with most new technology, various levels of automation will be demanded and introduced to the market to support needs in drug discovery and library generation. This technology will eventually replace hot plates, heating mantles, and block heaters, allowing chemists to begin using microwave energy on a broad scale, as affordable instrumentation becomes readily available. Academia, drug discovery, and lead optimization are the areas expected to receive the most benefit from this new technology. As microwave synthesis instrumentation continues to evolve, new applications will be developed for a variety of chemistries and process developing needs. This will naturally accelerate as the technology is adopted. Undoubtedly, microwave-enhanced synthesis will be a valuable tool for chemists in a variety of fields and specialties for many years to come.

Chapter 5
Getting Started With Microwave Synthesis

Are you ready to get started with microwave synthesis? If so, you're in the right place! This is probably the most important chapter in this book, for it provides the user with a quick review of previous chapters and an explanation of how to actually get started with microwave organic chemistry and perform reactions. As you gain more experience developing and performing reactions, you will be able to design, refine, and optimize your own methods. For a pictorial view, I have developed a flow chart that can be used to help follow this discussion (Chart 1). It has been divided into two parts (Charts 2 and 3). Chart 1 offers an overview of method development for closed and open vessel reactions. Chart 2 takes you through the development of a pressurized microwave reaction, while Chart 3 discusses one performed at atmospheric pressure. I will be referring to these charts throughout the chapter.

There is a completely new side of organic synthesis that is waiting to be discovered.

This chapter has been organized into two major sections. The first section pertains to method development. It is divided into three main parts, each discussing an important aspect of a typical, microwave-enhanced chemical reaction including atmospheric versus closed conditions; choosing a solvent; and deciding time, temperature, and power parameters. The last section discusses the optimization of your microwave reaction method. What happens if your first method did not work and no product has formed? I will provide details on which parameters you should change and how you should change them. Read on and let's get started!

Method Development

I. Pressure vs. Atmospheric

Once you have decided what type of chemical reaction you wish to perform, your first question is going to be whether you should run it in a closed environment or at atmospheric pressure. There are advantages to both. The scale of the reaction will probably be the deciding factor. Pressurized reactions are, of course, smaller in scale, as the certified pressure tubes can only hold about 7 mL. The maximum size is 10 mL, but there needs to be enough headspace to contain the vapors that result. As previously mentioned, a pressurized environment can be very advantageous to many different kinds of chemistries. Solvents can be heated up to temperatures that are two to four times their respective boiling points. Dichloromethane (bp 40 °C) can even be heated to 180 °C, which is 4.5 times its boiling point. This characteristic of microwave synthesis provides the large rate enhancements (up to 1000x) that are observed.

> ***Pressurized reactions provide inert atmospheres for use of air- and moisture-sensitive reagents.***

Chart 1

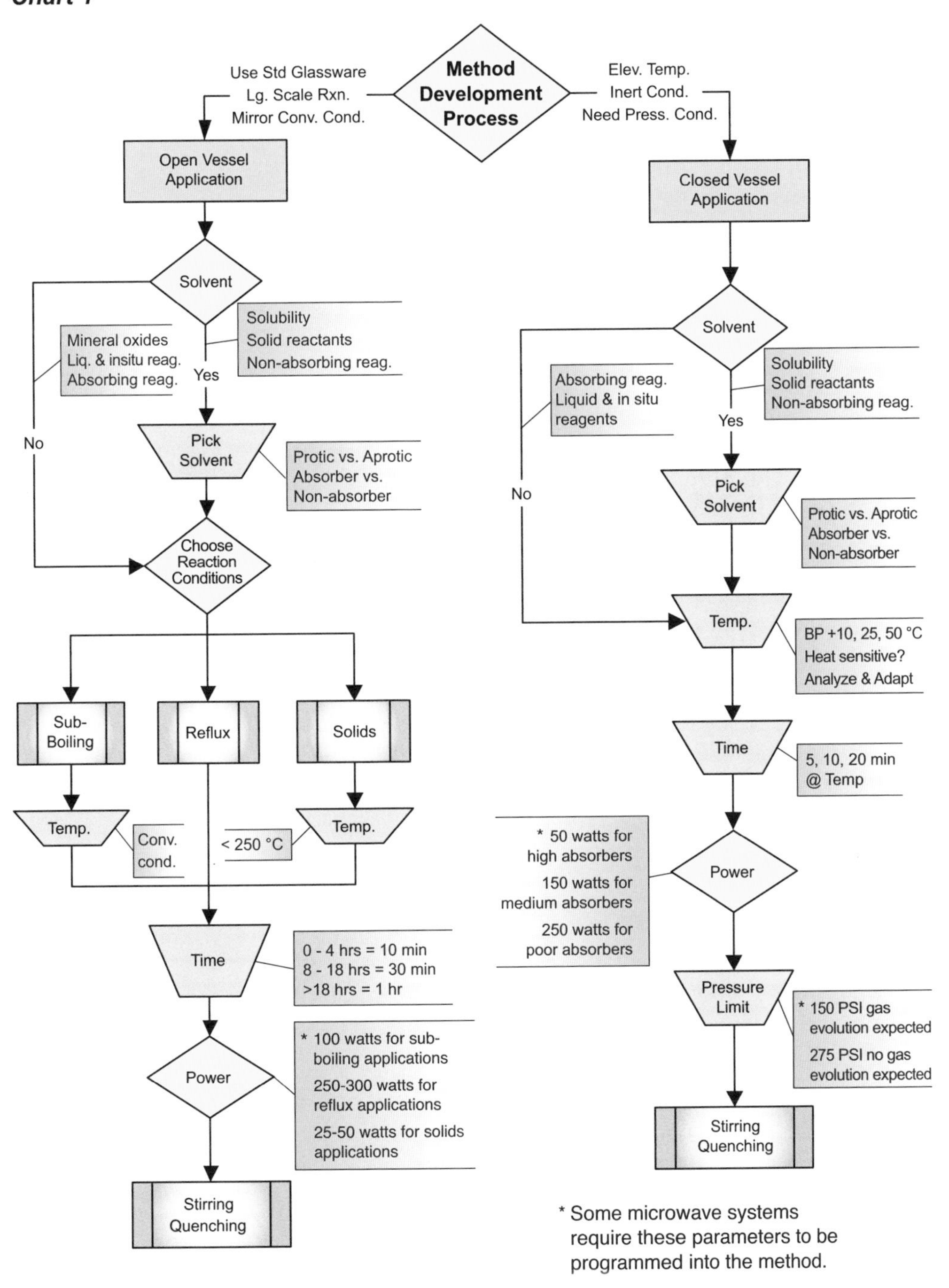

* Some microwave systems require these parameters to be programmed into the method.

Pressurized reactions also provide inert atmospheres for use of air- and moisture-sensitve reagents.

Atmospheric reactions can be performed on a much larger scale than pressurized ones. One major advantage is that they can be done in standard round bottom flasks. This allows for use of reflux condensers, addition funnels, Dean-Stark traps, or any other glass apparatus that is needed. Another major advantage of open vessel microwave organic reactions is that you can mirror the conventional conditions. Under atmospheric conditions, you may not observe the enhancements seen with pressurized vessels, but you will see reaction rates that are 10x faster than conventional methods. Depending on the type of reaction you are doing and the temperatures that are needed, atmospheric conditions may be the method of choice.

II. Choosing a solvent

As we move on in Chart 1, the next box we come to asks us whether we are going to use a solvent. Chapter 3 thoroughly discusses both reactions run in the presence of solvent and those run in a solventless environment. It is, once again, your preference, and there are benefits to each. Both open and closed vessel reactions can be performed either way. Most chemists are more familiar with solution phase reactions, but those performed in a solventless environment are becoming more prevalent in organic chemistry. An increasing need for less hazardous reaction conditions and environmentally safe procedures, or green chemistry, has led chemical synthesis in this direction.

There are other reasons to perform solvent-free microwave reactions. If all your reagents are in the liquid form (including in situ reagents) or your solid reagents melt at a certain temperature, then no additional liquid/solvent may be needed. In addition, your reaction mixture could be very "absorbing" without solvent. Polar or ionic reagents will couple very efficiently with the microwave energy. Most of the solvent-free research in which reagents are adsorbed onto mineral oxides has been performed in an open vessel. These reactions usually require larger vessels and mechanical stirring. Depending on what type of reaction you are attempting, a solvent-free method may be optimal.

For solution phase reactions, choice of solvent can be a crucial factor in the outcome. The polarity of a solvent plays a significant role in microwave-assisted reactions. If your reactants are "non-absorbing", then a polar solvent is necessary. The more polar a reaction mixture is, the greater its ability to couple with the microwave energy, leading to a more rapid rise in internal temperature. Table 28 (taken from Chapter 2) lists some common organic solvents that have been categorized as high, medium, or low absorbers of microwave energy. Though nonpolar solvents (i.e. hexane, benzene, toluene) do not couple very efficiently to microwave energy, and hence, do not heat reactions very well, they can act as a heat sink. Reaction mixtures that are temperature sensitive will benefit greatly from this ability, as the nonpolar solvent will help to draw away the thermal heat that is being produced from the interaction between microwave irradiation and the polar reagents. The reaction is still receiving activation energy, but its internal temperature will remain low.

Additionally, a pressurized environment can be very advantageous, as microwave energy (300 W) will reach and bypass the boiling point of most solvents in a matter of seconds. Certified pressure vessels allow for greater use of the lower boiling point solvents that are normally ignored in conventional high-temperature reactions. For specific information on individual solvents, Chapter 2 provides tables and figures of pressures generated at specific temperatures for different volumes of 25 common solvents.

Table 28

High, Medium, and Low absorbing solvents

Absorbance Level	Solvents
High	DMSO; EtOH; MeOH; Propanols; Nitrobenzene; Formic Acid; Ethylene Glycol
Medium	Water; DMF; NMP; Butanols; Acetonitrile; HMPA; Methyl Ethyl Ketone, Acetone, and other ketones; Nitromethane; *o*-Dichlorobenzene; 1,2-Dichloroethane; 2-Methoxyethanol; Acetic Acid; Trifluoroacetic Acid
Low	Chloroform; Dichloromethane; Carbon Tetrachloride; 1,4-Dioxane, THF, Glyme, and other ethers; Ethyl Acetate; Pyridine; Triethylamine; Toluene; Benzene; Chlorobenzene; Xylenes; Pentane, Hexane, and other hydrocarbons

General synthetic organic chemistry rules still apply with microwave-assisted chemical reactions. Regardless of the kind of reaction performed, the type of solvent for each remains the same. There are protic and aprotic solvents, and each of these may or may not be applicable for certain kinds of chemistry. Protic solvents have the ability to solvate or interact with both cations and anions, whereas aprotics can only solvate cations. The solvents of each type are interspersed throughout Table 28.

III. Temperature, Time, and Power

Once the solvent has been chosen or you have decided to go solvent-free, it is time to design the reaction method. There are three important variables to think about: temperature, irradiation time, and power. These parameters are presented in sequential boxes on the flow chart for each reaction type. They will vary with solvent selection and the choice of open or closed vessel conditions. We are all very familiar with temperature and time, as these dictate how we run conventional reactions. Power, however, is a new variable to consider in microwave-enhanced reactions. It is also probably the most important, but I will discuss this last.

As a traditional organic chemist, you are most concerned about the temperature of your reaction. In microwave reactions performed in pressure vessels (both with and without solvent), the best place to start is ten degrees above the temperature used in the conventional method. If you have chosen to do atmospheric work, follow the left side of Chart 1 according to whether or not you are using solvent. For solvent-free reactions (use of mineral oxides), you could start around 200 °C, *but I would not go above 250 °C, as these reaction mixtures will heat quickly.* For reactions in solvent, you will have to decide whether you are going to reflux or work with sub-boiling conditions. Set the temperature at least 50 degrees above the boiling point for reflux conditions. Solvents will reach temperatures that are 10-20 degrees above their boiling points in atmospheric microwave-assisted reactions. Setting a high temperature will also ensure a high, constant power level for direct molecular heating. In addition, remember to allow enough head space in your round-bottom flask for

Solvents will reach temperatures that are 10-20 degrees above their boiling points in atmospheric microwave-assisted reactions.

rapidly boiling reaction mixtures. It is also wise to use reflux condensers that are at least one foot in length, as solvents at temperatures above their boiling points will rapidly climb the height of the condenser. For sub-boiling temperatures, mimic the conventional method. Begin with the same temperature that you would normally use on the hotplate.

Deciding on how long to run a microwave reaction also depends on the type of reaction being performed. A good starting point for pressurized reactions (both with or without solvent) is 5-10 minutes. I would also use a 5-10 minute reaction time for reactions performed on mineral oxides. For solution phase atmospheric work, use the following reference chart to begin:

Conventional	Microwave
4 hrs	10 min
8-18 hrs	30 min
> 18 hrs	1 hr

The amount of power being applied to a microwave reaction is very important. Obviously, a low power level might not provide successful results, yet excessive power may cause decomposition. We already know that 300 W of microwave energy will reach and bypass the boiling point of most solvents in a matter of seconds, but do we always need this maximum power value with every reaction? The answer is no. Remember, organic reactions contain many different reagents and catalysts. Their presence can drastically enhance the coupling efficiency of a reaction mixture, regardless of solvent. In addition, many reagents and products are very sensitive to high temperatures and decompose readily. Applying a lower power for a selected amount of time at a certain temperature can sometimes be more effective.

Some microwave systems require the user to program the power parameter. For those that do, this paragraph is important for method development. In a closed reaction, a vessel failure can occur if the pressure rises too quickly because of abundant microwave energy. How do you know the right power level? With any new reaction, especially if you are unsure about how it will react in a microwave, you should start with 50 W. You will know instantly (ca. 5-10 sec) whether it is enough. If your reaction is struggling to reach its designated temperature, then you will have to increase the power. For open vessel, solvent-free reactions, I would start in the 25-50 W range. For refluxing under atmospheric conditions, this is one example where 250-300 W is necessary. A high power level will ensure that there is always constant microwave power being applied, and will keep your reaction mixture at its maximum attainable temperature. Finally, when mimicking conventional methods and working with sub-boiling temperatures, start with 100 W.

High energy is the reason microwaves are so beneficial to organic synthesis and why they have produced such dramatically favorable results. The energy transfer in a microwave-assisted reaction is incredibly quick, as energy is transferred every nanosecond it is applied. Conventional heating methods cannot do that. In a microwave reaction, as the temperature reaches the set value, the power is reduced so that the reaction mixture does not bypass that temperature point. It then stays at a lower level in order to maintain the set temperature throughout the entire reaction. The power, or energy, is the most important variable in a microwave-enhanced reaction. Recent experimentation has shown that simultaneous cooling of the reaction vessel during a reaction will ensure a constant, high power level for direct molecular

Simultaneous cooling of the reaction vessel during a reaction can dramatically improve the product yield of some reactions.

heating. *This has dramatically affected reaction rates and nearly doubled percent yields of some lower yielding reactions.*[18] Simultaneous cooling can also be especially useful in reactions where the reagents and/or products are heat sensitive. If compressed air is introduced to the cavity while simultaneously applying microwave irradiation, the thermal heat will not accumulate in the reaction mixture. Large amounts of energy can still be applied, while the bulk temperature remains low. This cooling feature can be applied to both closed and open vessel reactions. If your microwave system has this controlled cooling feature, I would highly suggest implementing it in every method.

Optimization

If the first method you programmed worked for you and you are happy with the results, then great! However, what happens if your first method did not work and no product has formed? Or, what if your product yield is very low? This is where optimization comes in. First, we'll discuss ways in which you can optimize your method. Then, as an example, I will optimize both the closed-vessel Negishi reaction (Scheme 141) that is shown on Chart 2 and the open-vessel nucleophilic aromatic substitution (S_NAr) reaction (Scheme 142) shown on Chart 3.

The first place that I would start when optimizing a closed-vessel method would be with the temperature parameter. As I previously mentioned, start ten degrees above the temperature used in the conventional method. Sequentially, increase the temperature to 25, 50, and 100 °C above the temperature in the conventional method. As long as you stay below the decomposition temperatures of heat-sensitive reagents, this should help you optimize your method. After finding the optimum temperature, you can then vary your reaction time to maximize product yield. Alternatively, if your

Chart 2

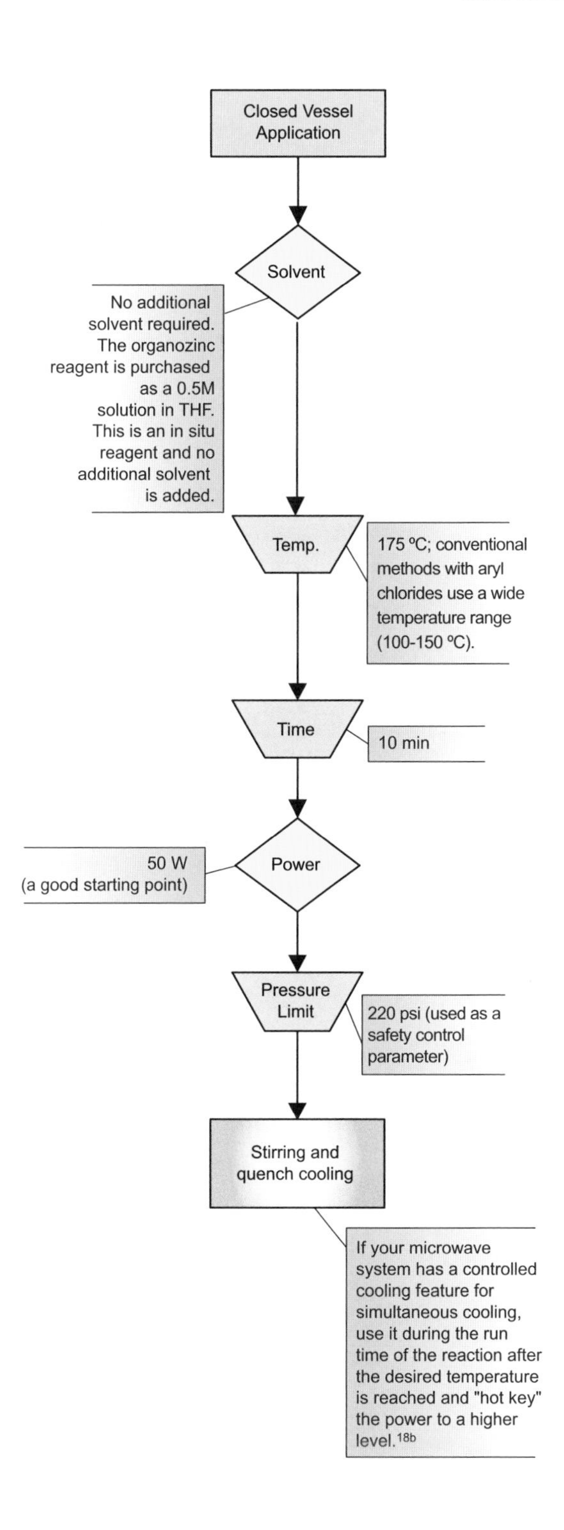

Closed Vessel Application
Solvent
No additional solvent required. The organozinc reagent is purchased as a 0.5M solution in THF. This is an in situ reagent and no additional solvent is added.
Temp.
175 ºC; conventional methods with aryl chlorides use a wide temperature range (100-150 ºC).
Time
10 min
Power
50 W (a good starting point)
Pressure Limit
220 psi (used as a safety control parameter)
Stirring and quench cooling
If your microwave system has a controlled cooling feature for simultaneous cooling, use it during the run time of the reaction after the desired temperature is reached and "hot key" the power to a higher level.[18b]

reaction is struggling to reach its designated temperature, then you will have to increase the power in small increments. If your microwave system has a controlled cooling feature for simultaneous cooling, use it during the run time of the reaction after the desired temperature is reached and "hot key" the power to a higher level.[18b] When optimizing reaction conditions and programming methods, remember to change only one variable at a time, so when something is successful, you will know which parameter it was.

When first attempting a closed-vessel, microwave-assisted Negishi reaction (Scheme 141, Chart 2), I examined the procedure of the conventional method. This method uses a wide range of thermal temperatures (100-150 °C), so I used 160 °C (ten degrees above the conventional method) as my starting point (Power: 50 W; Reaction time: 10 min). Upon analysis, no product was detected. I performed three additional runs at 175, 210, and 240 °C. Product was present in all three, but the percent yield decreased as temperature increased. As the temperature was increased, the predominant species was a 4,4'-dimethoxybiphenyl byproduct. From the results of this experiment, I realized that I should keep the temperature of my reactions around 175 °C.

Scheme 141

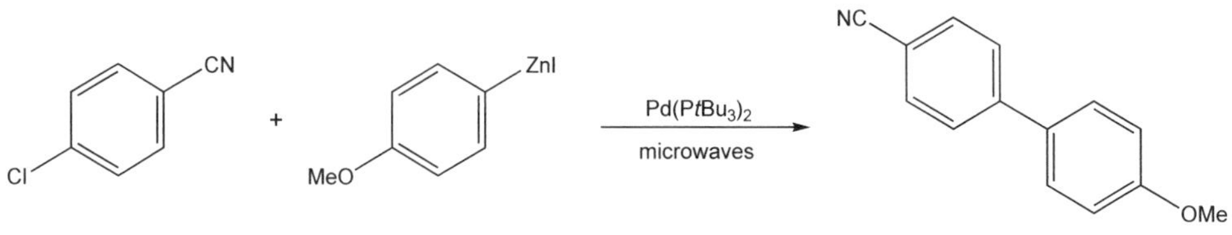

CN Cl + ZnI MeO —$Pd(PtBu_3)_2$, microwaves→ NC OMe

Once I figured out what temperature to run my Negishi reactions at, I then decided to experiment with reaction times. I ran a series of reactions at 175 °C (Power: 50 W) for 1, 5, 10, and 20 minutes. Upon analysis, I noticed considerable yield increases from one to five minutes and then from five to ten minutes. When I ran the reaction for 20 minutes, I did not notice any significant increase in product yield. In fact, the yield of the biphenyl byproduct increased with this extended reaction time. Thus, the results indicated that holding the reaction for ten minutes at 175 °C was optimal.

All my reactions up to this point were run with an initial power input of 50 W. The reaction mixture was very polar and the temperature point was reached quickly. Once the temperature was reached, the power level decreased. After I determined the optimal reaction temperature and run time, I wanted to ensure a high, constant power level. This can help struggling reactants acquire enough energy for transformation into products. The microwave system used with these reactions had the controlled cooling feature for simultaneous cooling. Once the desired temperature of 175 °C was reached, I administered the cooling feature and "hot keyed" the power up to 80 W in five-watt increments.[18b] In some Negishi reactions that were lower yielding, cooling helped immensely.

Reaction time is usually the main consideration when optimizing open-vessel microwave reactions. This is because the temperature of a particular reaction mixture can only go so high at atmospheric pressure. Conventionally, S_NAr reactions (Scheme 142, Chart 3) can require reflux conditions overnight, and sometimes, even 24 hours is needed. When I first attempted the reaction shown in the scheme, I chose to run it for 30 minutes. Equipping my instrument with a 1.5 foot reflux condenser, the following additional parameters were set: 250 W of power; 175 °C. Using acetonitrile (bp 82 °C) as my solvent, I set a high temperature to ensure a constant power level of 250 W. The temperature reached 95 °C after one minute and remained there for the duration of the reaction. Upon completion and analysis with GC/MS, I obtained my product in quantitative yield.

Chart 3

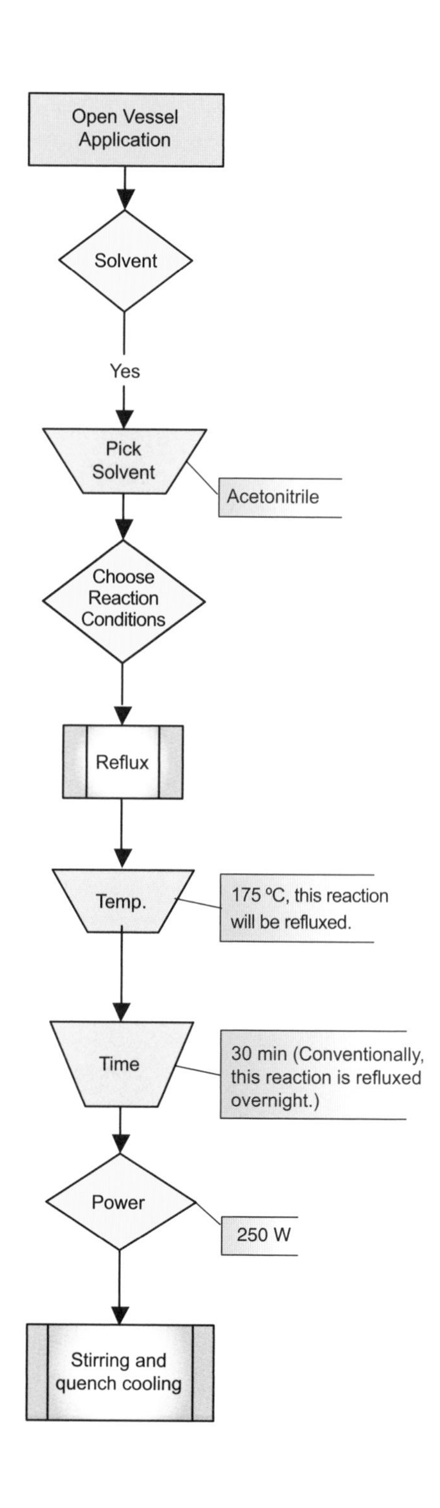
Open Vessel Application
Solvent
Yes
Pick Solvent
Acetonitrile
Choose Reaction Conditions
Reflux
Temp.
175 ºC, this reaction will be refluxed.
Time
30 min (Conventionally, this reaction is refluxed overnight.)
Power
250 W
Stirring and quench cooling

Scheme 142

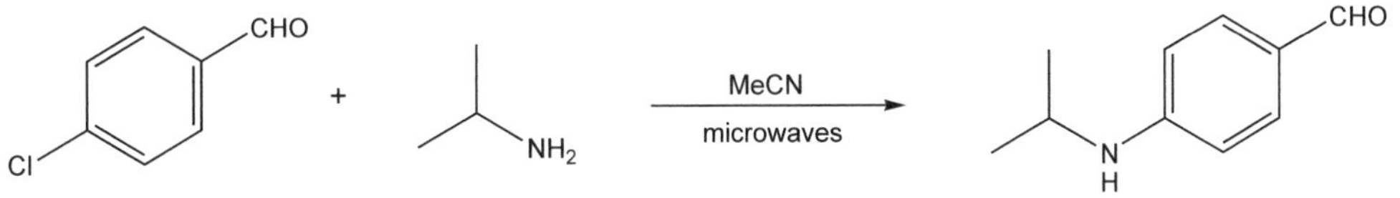

Let's assume that this reaction had not worked and the starting materials were the only compounds recovered, or very little product was obtained. The first change I would make is to increase the reaction time. If there is no product present after the initial run period, I would double the reaction time. In this case, run the reaction for an additional 30 minutes, and reanalyze the resulting mixture. In situations where the product is beginning to show, but in very small amounts, I would increase the reaction time in 10-minute increments.

If increasing the reaction time is not solving the problem, then temperature can be optimized. However, the only way to do this is by changing the solvent. If you want to increase the bulk temperature of your reaction, you must choose a solvent with a higher boiling point. If this is the case, I refer you back to the second part of the method development section in this chapter. There are many different solvents to choose from and the reaction chemistry you are performing will dictate this choice. You may also opt to use simultaneous cooling, if your microwave instrumentation is equipped with this feature. It will allow you to apply up to 300 W of microwave energy to your reaction system, but the bulk temperature will remain low. In the case of the S_NAr reaction, I would have been able to apply 300 W to my reaction, but the overall temperature would have been much lower than 95 °C. Once again, remember to change only one variable at a time, so when something is successful, you will know which parameter it was.

Though this chapter was written with the intent of providing the reader with basic knowledge on getting started using microwave instrumentation and some guidelines in designing and optimizing program methods, the best way to learn microwave synthesis is to actually do it. You cannot hurt the instrument, so be aggressive and innovative. Vessel failures are going to happen occasionally — it is OK! Just as you would do if a reaction flask exploded in your hood, pick up the pieces and move on. There is a completely new side of organic synthesis that is waiting to be discovered.

Chapter 6
Microwave Safety Considerations

Microwave safety may seem like a simple topic, but the rapid transfer of energy associated with microwave-enhanced chemistry does create safety issues. Is microwave irradiation safe? The answer is yes, but only with equipment that has been properly designed for its specific use. Closed vessel microwave systems have been in common use in laboratories since 1985. They have become the preferred method of sample preparation for many laboratory analyses. During that period, there have been documented cases of vessel failures. Reasons for these failures vary from exceeding the load limit of the vessel to using vessels well beyond their serviceable lifetime, or exceeding the pressure or temperature rating of the vessels. Frequently, such events are caused by the chemist being unfamiliar with the kinetics of the reaction. The best microwave safety device is a trained and knowledgeable operator. When performing microwave synthesis, a

The best microwave safety device is a trained and knowledgeable operator.

chemist should also be aware of the equipment being used and the stability of solvents at high temperatures. This chapter discusses these issues in hopes of increasing microwave safety awareness.

I. Equipment

Using the correct hardware for microwave synthesis is imperative for personal safety. DO NOT PURCHASE A DOMESTIC MICROWAVE OVEN FROM AN APPLIANCE STORE! It may seem to be a more economical solution, but these ovens are not designed for the rigors of laboratory usage. There are no safety controls or monitoring of power, temperature, or pressure. Acids and solvents corrode the interiors quickly, and the cavities are not designed to withstand the resulting explosive force from a vessel failure in runaway reactions. In addition, safety interlocks have been compromised allowing the unit to continue producing microwave energy even though the door has been opened.

> ***Using the correct hardware for microwave synthesis is imperative for personal safety.***

The majority of the research discussed in the applications chapter has been executed in multi-mode domestic microwave ovens. In the 1980s, laboratory instrument companies began to address the specific microwave safety issues. These instruments featured corrosion-resistant stainless steel cavities with reinforced doors. In the event of a vessel failure, the vessel and its contents would be contained within the cavity. Venting mechanisms were added for vapor accumulation in order to prevent potential explosions. Power, temperature, and pressure monitoring, with automatic safety controls, were also installed.

Single-mode microwave instrumentation is now available. These cavities are designed to provide a more consistent energy distribution with reproducible, stable energy patterns. The instruments are equipped with the same precautionary mechanisms as multi-mode cavities and both the temperature and pressure input values are used as safety parameters. Microwave power is automatically lowered just before either value is reached. Just as with any variable controller, power is cycled to maintain the operator-set parameter of pressure and/or temperature. Single-mode laboratory systems also act as a containment in the event of a vessel failure. The operator should always be sure to utilize the certified pressure tubes and accessories supplied by the original manufacturer. Placing any item inside a microwave cavity, which has not been designed, tested, and certified for use in that specific cavity, most assuredly will result in a failure of the equipment and/or the reaction.

The operator should always be sure to utilize the certified pressure tubes and accessories supplied by the original manufacturer.

II. Chemical applications and safety

Another important safety issue in microwave synthesis is the actual chemistry being performed. The chemist must be aware of the potential kinetics of the reaction to be accomplished. They should also be aware of the stability of their reagents at high temperatures. Many solvents and reagents decompose to hazardous components from prolonged exposure to high temperatures. This information is provided in Section 10 (Stability and Reactivity) of the Material Safety Data Sheet (MSDS) for each chemical. Potentially risky chemistries include those that are also unsafe under conventional heating conditions. Both azide and nitro groups have been known to cause explosions with thermal heat. Precautions should be taken when using microwave irradiation with compounds containing these functional groups. Additionally, any exothermic reaction should be treated carefully because of the fast energy transfer associated with microwave irradiation. An exothermic reaction is uncontrolled. It will only stop when the available fuels are expended. The production of pressure and heat happens at an alarmingly fast rate and can exceed the ability of the designed vent mechanisms on the vessel to safely relieve the condition. When this happens a vessel failure is imminent. A laboratory microwave system will contain the energy of the resulting failure. A well-designed system will not sustain damage. It will also be able to be cleaned and placed back into service in a matter of minutes.

A question that is always raised is whether transition metals can be used as catalysts in a microwave-assisted reaction. Absolutely! Microwave irradiation can greatly enhance organometallic reactions. When using a metal catalyst, only small amounts of ground material are needed, and this will not cause arcing within the microwave field. Conversely, metal filings and other ungrounded metals within the microwave field should be avoided, as they do provide a potential arc source.

Transition metals can be used as catalysts in microwave-assisted reactions.

Microwave chemists should also be aware of the potential for localized superheating. This can occur in a viscous sample when there is not proper stirring. When performing transition metal catalyzed reactions, a metallic coating on the vessel wall may result. The coating absorbs energy extremely well, heats quickly, and could melt the reaction tube. This can also occur in solvent-free reactions, especially when the reagents are adsorbed onto a mineral oxide. To reduce this occurrence, ensure adequate stirring with a heavier stir bar in pressurized reactions or with a mixer for open vessel, solvent-free experiments.

If you come away with one thing from this chapter, it should be to use equipment designed for the task and receive training on its utilization! Microwave irradiation provides more energy than thermal heat — there is no limit. If you are unsure about a particular reaction, then start small: use small amounts of reagents and start with a low power level and temperature. You can always increase the temperature or power level after observing the results. Work with chemicals in a laboratory hood to eliminate inhaling toxic fumes that can result from reagents and solvents exposed to high temperatures. Becoming familiar with microwave instrumentation and the hardware associated with it is also necessary. Used correctly, microwave technology provides a very safe way to perform chemistry, as the reaction vessel is more contained than when utilizing conventional heating methods.

Chapter 7
Microwave Hardware

The first microwave systems employed for laboratory applications were simple, inexpensive, domestic microwave ovens designed for home use. Creative chemists saw these devices as a cost-effective means to investigate the use of microwave energy to drive chemical reactions. These systems were commonly available, easy to operate, and due to their unsophisticated design, simple to modify for concept testing. For the most part, applications that required long heating cycles or had high temperature needs under conventional methods, such as sample preparation for trace analysis, were explored first.

Microwave instrumentation has evolved to include systems designed specifically to meet the needs of synthetic chemists.

Results from the use of microwave energy to promote these applications were often promising, but with safety and hardware limitations. Namely, these systems were not designed for the harsh conditions encountered in the laboratory, nor did they permit the kind of flexibility in reaction handling and software programming necessary to gain large scale use. The need for equipment specifically designed for the laboratory became more evident as experimentation with microwave-enhanced chemistry began to progress.

I. Multi-mode vs. Single-mode

The earliest attempts to design microwave-based laboratory instrumentation centered on the modification of domestic microwave ovens to enhance their ability to survive laboratory conditions. As such, multi-mode microwave applicators, the basic applicator used in domestic ovens, saw the earliest use in the laboratory. These applicators feature larger cavity geometries, which allow the processing of multiple samples simultaneously. Sample containers of various sizes and shapes from micro titer plates to large vessels, such as three L-flasks have been successfully used in multi-mode systems.

Multi-mode cavities, due to the physics of their design, have multiple pockets of energy dispersed throughout the cavity volume. Multiple energy pockets will have different levels of energy intensity, sometimes referred to as hot and cold spots. Therefore, any one sample may encounter an energy pocket of greater or lesser intensity than another if positioned statically in a multi-mode cavity. In order to provide an equal distribution of energy, multi-mode systems continuously rotate samples throughout the energy field. This tends to smooth or average the field exposure across all of the samples during the energy cycle.

For the organic chemist, microwave systems seem to provide an obvious resolution for the need to accelerate long, laborious reaction procedures. Medicinal chemists see microwave systems as a way to accelerate optimization protocols. However, it is the combinatorial chemist who greets the use of microwave chemistry most openly — as a way to increase reaction throughput in parallel chemistry applications. In all cases, the use of multi-mode microwave systems has proven to be of limited advantage.

For organic and medicinal chemists, the lack of field homogeneity paired with the positional sensitivity of these systems produce results that are at once promising and distressing. Often reactions will show great improvements only to have reproduction of the result elude the investigator. For the combinatorial chemist, the combination of field inhomogeneity coupled with the differences in sample absorption characteristics make the control of chemistries conducted simultaneously in parallel formats difficult when diverse chemistries are attempted. Specifically, multi-mode systems become problematic for chemists trying to reproduce their reactions on a small scale. While the total power generated may be high in industrial multi-mode instruments (typically in the 1000 – 1200 W range), the power density of the field is quite low due to the total volume of the cavity (typically in the 0.025 – 0.040 W/mL range). Therefore, trying to heat the small individual samples characteristic of drug discovery or new chemistry research is difficult.

To combat the inherent difficulty with multi-mode technology, instrumentation manufacturers developed single-mode cavities with more consistent and predictable energy distribution.[13] Single-mode instruments produce one homogenous, intense pocket of energy that is highly reproducible. Due to their uniform energy distribution and higher power density, these systems typically couple more efficiently with small samples. Although single-mode systems only output 300 – 400 W of power, their smaller cavity volume and single, focused energy pocket yield a field density in the 0.90 W/mL range.

Single-mode systems were not widely used, however, until recent advances in technology led to the development of instruments with the necessary features and software to be of use in organic synthesis laboratories. Chemists wanted the ability to perform elevated pressure and atmospheric reactions. In addition, they needed a way to "tune" the instrument for particular applications, as changes in sample size or the physical coupling characteristics of the sample (i.e. polarity of the sample, conductive properties of the sample) can dramatically affect the ability of the applicator to couple with the sample. Only in the last few years have systems become available that either allow the user to tune the applicator or provide applicators that automatically tune to changing application needs.

The need to change the tuning of these cavities can be difficult if a chemist is processing several different types of samples concurrently. Figure 43 shows the side view of an older single-mode design. There is a rectangular waveguide (the microwave cavity), a power source, a sample

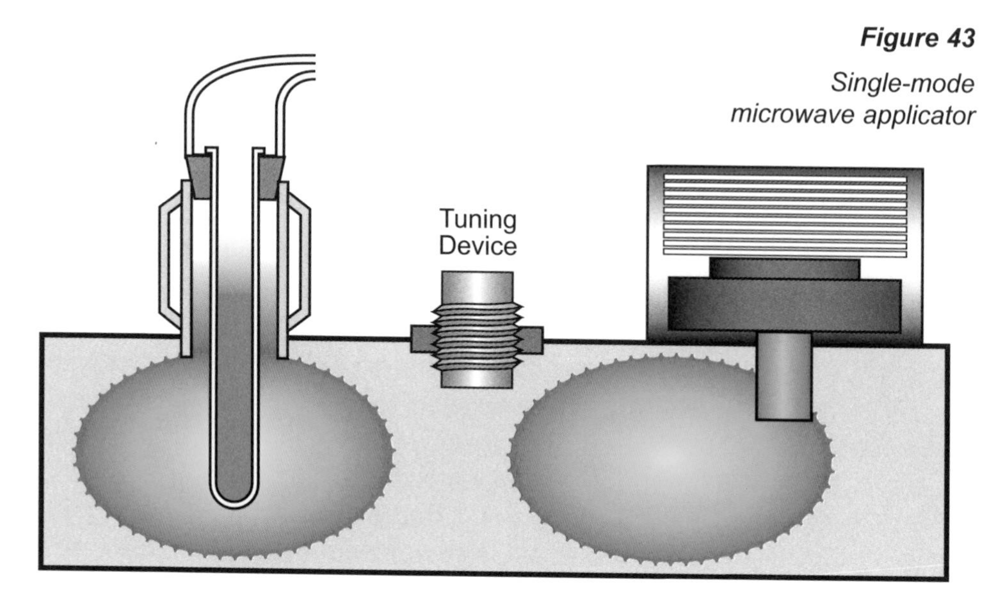

Figure 43

Single-mode microwave applicator

positioned at a maximized energy point from the magnetron, and some type of mechanical tuning device in the system that will adjust it for variations in the sample.[8,12,722,723]

Due to recent advances in single-mode microwave technology, systems now offer greater flexibility to the organic chemist. In addition to the traditional rectangular waveguide style applicator, a circular waveguide capable of self-tuning is now available (Figure 44, top view). This applicator features multiple entry points for the microwave energy to enter the cavity, compensating for variations in the coupling characteristics of the sample, the physical size of the sample, and the geometrical placement of the sample in the cavity. This design feature effectively renders the cavity immune to tuning issues.

The circular waveguide design also equates to a larger volume than preceding single-mode designs, offering flexibility in terms of sample volume. It can accept sample containers ranging from 1 mL up to 125 mL in size. Using infrared thermometry technology, this system allows volumes as low as 100 μL in specially designed pressure tubes. Additionally, it can be used with open vessels for performing traditional atmospheric work. This provides an open system in which the

Figure 44

Self-tuning single-mode microwave applicator

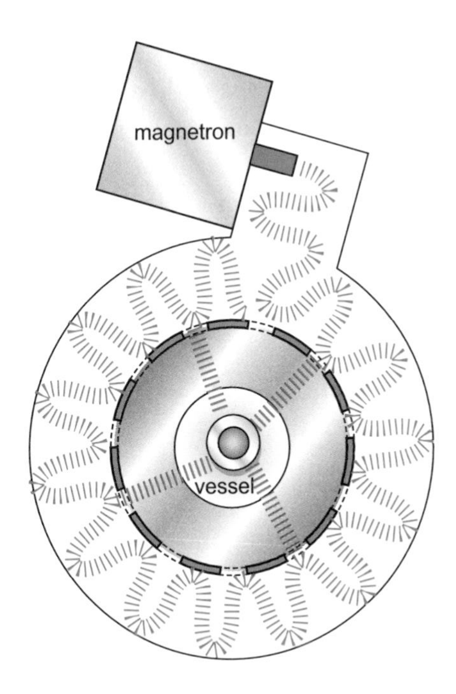

chemist can add reagents to the sample while it refluxes. This technology has incorporated all of the desirable capabilities of conventional heating methods into a microwave system while also offering the benefits of greatly increased reaction times and improved yields.

III. Pressurized and Atmospheric hardware

Most single-mode applicators permit the use of sealed vials to perform elevated pressure reactions. As previously mentioned, the circular waveguide design can also perform reactions at atmospheric pressure. This applicator design employs two different sized attenuators (or doors), one to use with pressure tubes and another that will fit around the necks of round-bottom flasks. This configuration allows the operator to choose the glassware to fit a specific application need.

Most systems incorporate the use of 10-mL pressure vials and are preferably capped with self-sealing septa. These systems monitor the reaction conditions, and offer feedback control of the power input based on this monitored information. The available systems offer two pressure sensor design philosophies, direct and indirect measurement. Direct pressure measurement inserts a needle probe into the septum of the reaction tube. It has the benefit of fast response time, allowing a precise control of the reaction environment. The disadvantage to the direct method is that the pressure in the vessel must be lowered before the cover may be removed. This typically takes about one minute from the completion of a run. With the indirect method, the deflection of the septum is measured and related to the internal pressure of the vial. This is beneficial when working with highly corrosive reagents, as there is no septum penetration, but it does present some lag time in the measurement.

Using the circular waveguide design, atmospheric reactions can be done in standard round-bottom flasks up to 125 mL in size and with neck joint sizes of 24/40 and smaller. Reflux condensers, addition funnels, Dean-Stark traps, or any other glass apparatus that is needed can be used. Depending on the type of reaction you are doing and the temperatures that are needed, atmospheric conditions may be the method of choice. Microwave organic reactions performed under atmospheric conditions may not produce the enhancements seen with pressurized vessels, but they can give reaction rates that are 10x faster than conventional methods and have the advantage of offering chemists the opportunity to increase the scale of the reaction.

IV. Cooling

Microwave irradiation can take reactions up to extremely high temperatures. Some single-mode systems include a cooling feature that allows for fast reaction quenching as compressed gas is forced into the reaction cavity. As the gas expands, it cools the atmosphere inside the cavity. The accelerated cooling profile reduces side reactions and provides for cleaner chemistries, resulting in less post-reaction workup (Figure 45).

Figure 45

Cooling performance

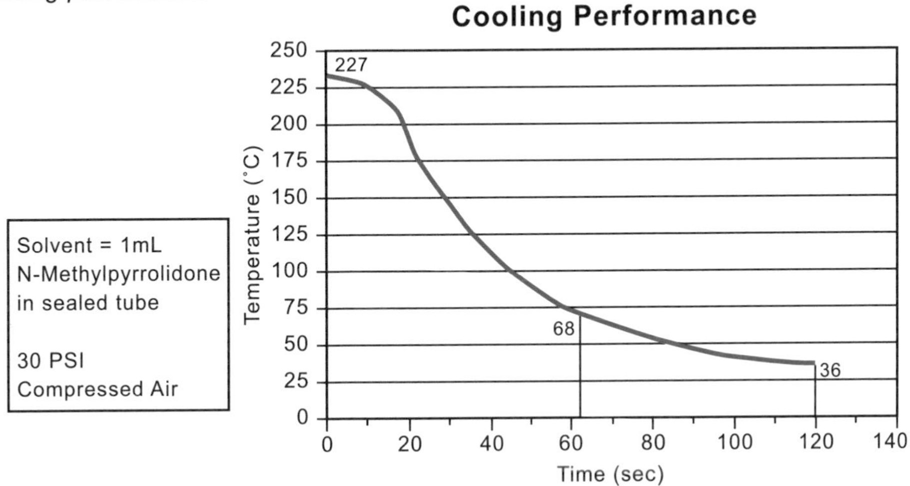

Some systems also have the ability to activate the cooling feature during reactions to control bulk temperature rise. This can be especially useful in reactions where the reagents and/or products are heat sensitive. If compressed air is introduced into the cavity while simultaneously applying microwave irradiation, thermal heat will not accumulate in the reaction mixture. Large amounts of energy can still be applied while the bulk temperature remains low, resulting in higher product yields and cleaner chemistries for many reactions.

V. Automation

Some single-mode systems are automated for increased throughput and unattended operation. As the typical reaction times are short (2-10 minutes), an automated sequential procedure allows for total reaction control and optimization while approaching parallel throughput results. The practice of High Throughput Sequential Chemistry (HTSC) allows for the ultimate in reaction optimization as the individual parameters of each reaction may be modified during the process to promote the most advantageous result.

As opposed to traditional parallel synthesis, where the reaction conditions are held constant for a parallel batch of chemistries, HTSC offers the chemist complete control over every step of each reaction in an optimization procedure, a feat that was too time-consuming and labor-intensive to be realistically attainable before. Automated formats permit the best practice of HTSC, as an entire library may be generated unattended with completely different reaction protocols within the same timeframe as traditional parallel schemes. Increased diversity and control without limited throughput make automated systems a popular choice with many chemists.

Afterword
Future Trends

At present, microwave synthesis is a fledgling science. However, its acceptance and evolution are progressing at a furious rate. Thus, it can be assumed that this shift to the application of microwave synthesis will only increase in the next few years. Instrument manufacturers are teaming with academic and industrial trailblazers in an effort to design and build affordable, flexible instrumentation. Hands-on application of these devices has resulted in and will continue to encourage innovative advances in the instruments, as well as an expanding applications and knowledge base. These factors will contribute to the rapid expansion, evolution and adoption of the technology.

Advantages of Microwave Synthesis

- *faster reactions*
- *higher yields*
- *cleaner reactions*
- *new pathways*
- *green chemistry*

I. Expanding use and range of applications

Considered exotic and out of place, microwave instruments were first used in the analytical laboratory only for the most difficult sample types. Today, microwave energy is a broad-based replacement for conventional heating methods. Standard analytical methods are developed and written around microwave laboratory equipment. Much as the analytical laboratory embraced microwave instruments in the mid-1980s, the pharmaceutical lab is beginning to use them as the primary and preferred way to perform chemical synthesis. Most of the major pharmaceutical companies are currently evaluating, or have already begun to shift toward, microwave synthesis for their drug discovery and development efforts. Should the current trend of adoption continue, in just over 3 years, as many as 50% of all medicinal chemists will be using microwave synthesis for most of their reactions. Biotechnology companies and academic institutions are also rapidly adopting microwave irradiation in their synthetic projects. The increasing number of publications on microwave synthesis is further evidence that the technology is rapidly becoming an accepted methodology within the synthetic community.

The speed and simplicity of microwave synthesis provides exciting new opportunities for performing laboratory experiments. Historically, most of the academic work has been done in Europe, but this is changing as a number of the major synthetic groups in the United States are now starting to use microwave synthesis in their research. As often happens with important, new technology, universities with strong science programs are beginning to include it in graduate research and undergraduate teaching programs to expose students to microwave synthesis as early as possible. One major university in the United States is currently investigating the use of microwave instrumentation in their undergraduate teaching curriculum and expects to initiate a pilot program within the next year.

II. Proteomic and genomic applications

Many expect microwave synthesis to find major applications in proteomics and genomics. As described and demonstrated in Chapter 5, the ability to simultaneously cool a reaction while applying microwave energy to the reactants offers the possibility of synthesizing large protein molecules. These molecules are very temperature sensitive. Typically, reaction times are quite long to insure the integrity of the molecules. Microwave energy rapidly drives these reactions through direct molecular heating. When coupled with cooling the bulk temperature of the reactants is kept low. Even with substantial microwave energy input (25 - 100 W, 6 - 24 cal/sec), reaction temperatures can be held between 30 - 40 °C.

Although not yet demonstrated, microwave synthesis also has the potential to be useful in the PCR (polymerase chain reaction) process, used for amplifying DNA fragments. Thermal cycling with conductive heat could be replaced with a microwave "electronic cycling" process. The microwave energy can easily be cycled in milliseconds, which — when combined with continuous cooling — could reduce cycle times by 100- to 1000-fold (minutes to milliseconds).

III. Scaling up reactions with flow-through systems

One application that is on the immediate horizon for microwave technology is flow-through synthesis. This allows for the constant reaction of the components, and therefore, the continuous on-line production of material. Flow-through systems based on multi-mode technology have had some success in the past in other analytical application areas. Typically, the technology has proven sound, but its deployment has failed to meet the needs of the synthetic chemist. Single-mode applicators with flow cells are proving to be a better match of technology with user need, because they have the ability to provide production amounts of material in smaller, simpler devices. Capital outlays and safety concerns are both minimized, and switching over from one process to another is cleaner, quicker, and less waste intensive. Single-mode-based flow module testing is underway at several industrial sites where microwave instrumentation suppliers have partnered with industrial producers to define platform options.

The use of microwave technology in this manner will offer a rapid means of production for bulk amounts of material. Flow-through systems will provide the pharmaceutical laboratory with a methodology to produce large quantities of final products. As a result, the process chemist will have access to all of the enhancements of microwave synthesis without having to forfeit the scale of material production needed to supply the marketplace.

Microwave synthesis is a breakthrough technology for chemistry: an idea whose time has come. Its use has become more widespread, and as the technology continues to rapidly evolve, microwave synthesis will have a dramatic impact on the world of chemistry.

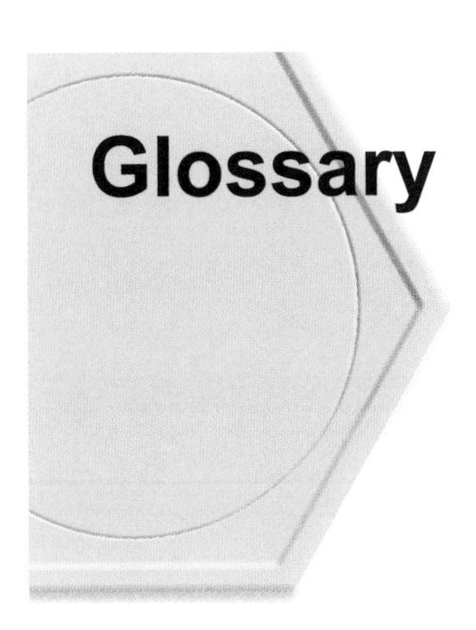

Glossary

Activation energy – The energy barrier that must be overcome in order for a chemical transformation to occur. Higher activation energy transformations, which are difficult or impossible to complete with conventional heating, can be done with microwave irradiation because the energy transfer is more efficient.

Arrhenius equation ($k = Ae^{-E_a/RT}$) – The relationship between the rate constant, the activation energy, and the temperature of a particular reaction. The rate constant (k) is dependent on two factors: the frequency of collisions between molecules that have the correct geometry for a reaction to occur (A) and the fraction of those molecules that have the minimum energy required to overcome the activation energy barrier ($e^{-E_a/RT}$). Microwave irradiation provides enhanced reaction rates (and rate constants), which are a result of the higher instantaneous temperatures caused by the rapid energy transfer.

Atmospheric reactions (open-vessel) - Reactions performed in standard glassware at atmospheric pressure. The bulk temperature is limited to 10-20 degrees above the atmospheric boiling point of the solvent. Microwave irradiation can enhance the reaction rates of these reactions by up to 10-fold. They can be equipped with reflux condensers, Dean-Stark traps, addition funnels, or any other glass apparatus that is needed.

Attenuator – A mechanical device used to block the passage of microwave irradiation. In some laboratory microwave systems, it is a cylindrical opening that allows access to the microwave chamber. The height of the cylinder must be 2.5 times the diameter to provide a sufficient barrier.

Conductive heating – Traditional method to perform thermally-assisted chemical reactions. Heat is applied externally and must first pass through the walls of the reaction vessel and the solvent. It is characterized as being slow and inefficient.

Cool reactions – The term used to describe microwave reactions performed at lower temperatures. The use of transparent (nonpolar) solvents and/or simultaneous cooling helps to minimize bulk heating. Recent experimentation has shown that simultaneous cooling of the reaction vessel during a reaction will ensure a constant, high power level for direct molecular heating. It also allows efficient transformations of larger, heat sensitive compounds (e.g. proteins) without causing sample degradation.

Dielectric constant (ε') – Also known as relative permittivity, it is the measure of a substances ability to store electric charges. It is dependent on both temperature and frequency.

Dielectric loss (ε") – Also known as complexed permittivity. The amount of input microwave energy that is lost to the sample by being dissipated as heat. It is this value that best provides the organic chemist with the coupling efficiency of a particular solvent.

Dipole moment – The measure of the permanent polarity of a molecule.

IMS frequencies – Four microwave frequencies that have been reserved for industrial, medical, and scientific applications. These frequencies are 915, 2450, 5800, and 22,125 MHz. The 2450 MHz frequency is currently the only one used for laboratory synthesis.

Kinetic product – The product that has a higher rate of formation because of lower activation energy. This is generally the preferred product at lower temperatures.

Magnetron – An electromagnetic device that generates microwaves at a fixed frequency. The most common frequency is 2450 MHz, which is currently used in all laboratory microwave instrumentation.

Microwave coupling – The direct transfer of microwave energy to a substance, resulting in instantaneous heating.

Microwave heating – Direct energy transfer from the microwaves to the interacting substances. This causes kinetic excitation, which results in chemical transformations and secondary heating. It is characterized by rapid and efficient energy transfer.

Microwave irradiation – A form of electromagnetic energy that falls in the frequency range of 0.3 to 300 GHz. It is a low energy, non-ionizing radiation that transfers energy by interacting with polar substances.

Molecular heating – The term used to describe the direct energy transfer from microwaves to the molecules being heated. The speed and efficiency of this energy transfer is the reason for the greatly enhanced chemical transformations seen with microwave irradiation.

Multi-mode cavity – Large chambers that are used in domestic microwave ovens and various laboratory systems. They are large enough to propagate multiple modes of microwave energy (typically 20 - 30), which interact with each other constructively and destructively and create "hot spots" and "cold spots". Uniform heating typically requires sample rotation throughout the energy field.

Nonpolar solvents – Solvents with small dipole moments that heat poorly or not at all with applied microwave energy. They are considered low absorbers and have dielectric losses (ε'') < 1.0.

Polar solvents – Solvents with large dipole moments that heat efficiently with applied microwave energy. They are considered medium or high absorbers and have dielectric losses (ε'') between 1.0 - 50.

Pressurized reactions (closed-vessel) – Reactions performed in 10-mL reaction tubes that are capped with self-sealing septa. This allows reagents and solvents to be heated above their normal boiling points. Microwave irradiation is ideal for performing pressurized reactions because of the rapid energy transfer and instantaneous heating that occurs. Chemical reaction rates can be enhanced by as much as 1,000-fold.

Reaction intermediates – Reactive species that are temporarily formed in chemical transformations. These intermediates are usually very polar and can interact directly with microwaves if they are long-lived ($> 10^{-9}$ seconds).

Reaction rate – The speed at which reactants are transformed into product(s). It is a function of temperature and concentration of reactants. The rapid energy transfer and high instantaneous temperatures from microwave irradiation can increase reaction rates 10 - 1,000-fold.

Single-mode cavity – A small chamber of various geometries (circular, rectangular, etc.) that is sized such that it will only propagate one mode of microwave energy. This creates a more homogenous energy distribution and a much higher power density than multimode cavities.

Tangent delta (δ) - Also known as loss tangent, it is the ratio of the dielectric loss (ε'') to the dielectric constant (ε') [$\tan \delta = \varepsilon''/\varepsilon'$]. It is a measure of the ability of a substance to convert electromagnetic energy into heat at a given frequency and temperature.

Thermodynamic product – The more stable reaction product. Formation of this product requires a larger activation energy and higher temperatures. Microwaves can cause a shift in product from kinetic to thermodynamic because of the more efficient energy transfer and high instantaneous temperatures.

Transition states – When two or more reactants come together with sufficient energy to overcome the activation barrier and exist in a transitory state prior to completion of the transformation. The lifetime of these transition states depends on their stability and can vary from 10^{-13} seconds (highly activated) to several seconds (resonance-stabilized). Since microwave energy is transferred at a rate of 10^{-9} seconds, it can directly interact with transition states that are longer-lived.

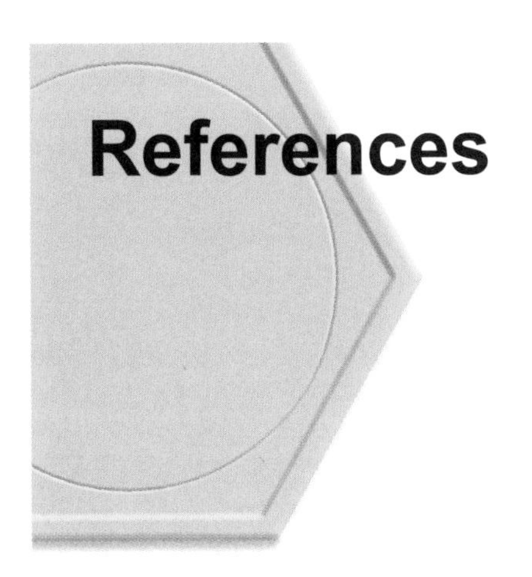

References

1. Neas, E.D.; Collins, M.J. *Introduction to Microwave Sample Preparation Theory and Practice*, Kingston, H.M.; Jassie, L.B., Eds., American Chemical Society **1988**, ch. 2, pp. 7-32.

2. Mingos, D.M.P.; Baghurst, D.R. *Microwave-Enhanced Chemistry Fundamentals, Sample Preparation, and Applications*, Kingston, H.M.; Haswell, S.J., Eds., American Chemical Society **1997**, ch. 1, pp. 3-53.

3. Giguere, R.J.; Bray, T.L.; Duncan, S.M.; Majetich, G. Application of commercial microwave ovens to organic synthesis. *Tetrahedron Lett.* **1986**, *27*, pp. 4945-48.

4. Gedye, R.; Smith, F.; Westaway, K.; Ali, H.; Baldisera, L.; Laberge, L.; Rousell, J. The use of microwave ovens for rapid organic synthesis. *Tetrahedron Lett.* **1986**, *27*, pp. 279-82.

5. Jun, C.H.; Chung, J.H.; Lee, D.Y.; Loupy, A.; Chatti, S. "Solvent-free chelation-assisted intermolecular hydro-acylation: effect of microwave irradiation in the synthesis of ketone from aldehyde and 1-alkene by Rh(I) complex." *Tetrahedron Lett.* **2001**, *42*, pp. 4803-05.

6. Loupy, A.; Perreus, L.; Liagre, M.; Burle, K.; Moneuse, M. "Reactivity and selectivity under microwaves in organic chemistry. Relation with medium effects and reaction mechanisms." *Pure Appl. Chem.* **2001**, *73*, pp. 161-66.

7. Mingos, D.M.P.; Baghurst, D.R. "Applications of microwave dielectric heating effects to synthetic problems in chemistry." *Chem. Soc. Rev.* **1991**, *20*, pp. 1-47.

8. Loupy, A.; Petit, A.; Hamelin, J.; Texier-Boullet, F.; Jacquault, P.; Mathe, D. "New solvent-free organic synthesis using focused microwaves." *Synthesis* **1998**, *9*, pp. 1213-34.

9. Loupy, A. "Microwaves in organic synthesis: a clean and high-performance methodology." *Spectra Anal.* **1993**, *22*, p. 175.

10. Majetich, G.; Hicks, R. "Applications of microwave-accelerated organic synthesis." *Radiat. Phys. Chem.* **1995**, *45*, pp. 567-79.

11. Bose, A.K.; Manhas, M.S.; Ghosh, M.; Shah, M.; Raju, V.S.; Bari, S.S.; Newaz, S.N.; Banik, B.K.; Chaudhary, A.G.; Barakat, K.J. "Microwave-induced organic reaction enhancement chemistry. 2. Simplified techniques." *J. Org. Chem.* **1991**, *56*, pp. 6968-70.

12. Johannsson, H. "A solution to the bottleneck in drug discovery." *Am. Laboratory* **2001**, *33*, pp. 28-32.

13. Larhed, M.; Hallberg, A. "Microwave-assisted high-speed chemistry: a new technique in drug discovery." *Drug Discovery Today* **2001**, *6*, pp. 406-16.

14. Strauss, C.R.; Trainor, R.W. "Developments in microwave-assisted organic chemistry." *Aust. J. Chem.* **1995**, *48*, pp. 1665-92.

15. Langa, F.; De La Cruz, P.; De La Hoz, A.; Diaz-Ortiz, A.; Diez-Barra, E. "Microwave irradiation: more than just a method for accelerating reactions." *Contemp. Org. Synth.* **1997**, *4*, pp. 373-86.

16. Kuhnert, N.; Danks, T.N. "Highly diastereoselective synthesis of 1,3-ozazolidines under thermodynamic control using focused microwave irradiation under solvent-free conditions." *Green Chem.* **2001**, *3*, pp. 68-70.

17. Gabriel, C.; Gabriel, S.; Grant, E.H.; Halstead, B.S.J.; Mingos, D.M.P. "Dielectric parameters relevant to microwave dielectric heating." *Chem. Soc. Rev.* **1998**, *27*, pp. 213-23.

18. a) Internal Communication, CEM Corporation, Matthews, NC. b) This reaction was performed on a CEM Discover System. This microwave system is equipped with "hot keys", which allow the user to change parameter values "on-the-fly". One of these hot keys is for simultaneous controlled cooling. Once the reaction has reached its programmed temperature, the cooling option can be turned on with the hot key. This will ensure a constant, high power level. Use of the power hot key can then allow the user to increase the power level in small increments in order to maximize the amount of microwave energy being delivered to the reaction.

19. Welton, T. "Room-temperature ionic liquids. Solvents for synthesis and catalysis." *Chem. Rev.* **1999**, *99*, pp. 2071-83.

20. a) Varma, R.S.; Namboodiri, V.V. "An expeditious solvent-free route to ionic liquids using microwaves." *J. Chem. Soc., Chem. Commun.* **2001**, *7*, pp. 643-44. b) Varma, R.S.; Namboodiri, V.V. "Solvent-free preparation of ionic liquids using a household microwave oven." *Pure Appl. Chem.* **2001**, *73*, pp. 1309-14.

21. Khadilkar, B.M.; Rebeiro, G.L. "Microwave assisted synthesis of room temperature ionic liquid precursor quaternary salts." *Fifth International Electronic Conference on Synthetic Organic Chemistry (ECSOC-5)* **2001**, E0020 (www.mdpi.net).

22. Leadbeater, N.E.; Torenius, H.M. "A study of the ionic liquid mediated microwave heating of organic solvents." *J. Org. Chem.* **2002**, *67*, pp. 3145-48.

23. Fraga-Dubreuil, J.; Bazureau, J.P. "Rate accelerations of 1,3-dipolar cycloaddition reactions in ionic liquids." *Tetrahedron Lett.* **2000**, *41*, pp. 7351-55.

24. Fraga-Dubreuil, J.; Bazureau, J.P. "Grafted ionic liquid-phase-supported synthesis of small organic molecules." *Tetrahedron Lett.* **2001**, *42*, pp. 6097-100.

25. Fraga-Dubreuil, J.; Bazureau, J.P. "Ionic liquid-phase organic synthesis (IoLiPOS) methodology assisted by focused microwave technology: first results." *Fifth International Electronic Conference on Synthetic Organic Chemistry (ECSOC-5)* **2001**, E0011 (www.mdpi.net).

26. March, J. *Advanced Organic Chemistry: Reactions, Mechanisms, and Structure*, 4th Ed., John Wiley and Sons, **1992**, ch. 10, p. 293, 340.

27. Yu, H.M.; Chen, S.T.; Chiou, S.H.; Wang, K.T. "Determination of amino acids on Merrifield resin by microwave hydrolysis." *J. Chromatogr.* **1988**, *456*, pp. 357-62.

28. Yu, H.M.; Chen, S.T.; Wang, K.T. "Enhanced coupling efficiency in solid-phase peptide synthesis by microwave irradiation." *J. Org. Chem.* **1992**, *57*, pp. 4781-84.

29. Strohmeier, G.A.; Kappe, C.O. "Rapid parallel synthesis of polymer-bound enones utilizing microwave-assisted solid-phase chemistry." *J. Comb. Chem.* **2002**, *4*, in press.

30. Lew, A.; Krutzik, P.O.; Hart, M.E.; Chamberlin, A.R. "Increasing rates of reaction: microwave-assisted organic synthesis for combinatorial chemistry." *J. Comb. Chem.* **2002**, *4*, in press.

31. Glass, B.M.; Combs, A.P. "Rapid parallel synthesis utilizing microwave irradiation." *Fifth International Electronic Conference on Synthetic Organic Chemistry (ECSOC-5)* **2001**, E0027 (www.mdpi.net).

32. Kaboudin, B.; Balakrishna, M.S. "Surface-mediated solid phase reactions: microwave assisted Arbuzov rearrangement on the solid surface." *Synth. Commun.* **2001**, *31*, pp. 2773-76.

33. Kaiser, N.F.K.; Hallberg, A.; Larhed, M. "Solid phase carbonylation in fast microwave-mediated combinatorial drug discovery." *Fifth International Electronic Conference on Synthetic Organic Chemistry (ECSOC-5)* **2001**, E0007 (www.mdpi.net).

34. Kappe, C.O. "Speeding up solid-phase chemistry by microwave irradiation: a tool for high-throughput synthesis." *Amer. Lab.* **2001**, *33*, pp. 13-19.

35. Stadler, A.; Kappe, C.O. "High-speed couplings and cleavages in microwave-heated, solid-phase reactions at high temperatures." *Eur. J. Org. Chem.* **2001**, pp. 919-25.

36. Hajipour, A.R.; Mallakpour, S.E.; Afrousheh, A. "One-pot and simple reaction for the synthesis of alkyl *p*-toluenesulfinate esters under solid-phase conditions." *Phosphorus Sulfur Silicon Relat. Elem.* **2000**, *160*, pp. 67-75.

37. Kidwai, M.; Misra, P.; Bhushan, K.R.; Saxena, R.K.; Singh, M. "Microwave-assisted solid-phase synthesis of cephalosporin derivatives with antibacterial activity." *Monatsh. Chem.* **2000**, *131*, pp. 937-43.

38. Kumar, H.M.S.; Anjaneyulu, S.; Reddy, B.V.S.; Yadav, J.S. "Microwave-assisted rapid Claisen rearrangement on solid phase." *Synlett.* **2000**, *8*, pp. 1129-30.

39. Stadler, A.; Kappe, C.O. "Solid phase coupling of benzoic acid to Wang resin: a comparison of thermal versus microwave heating." *Fourth International Electronic Conference on Synthetic Organic Chemistry (ECSOC-4)* **2000**, B0002 (www.mdpi.net).

40. Scharn, D.; Wenschuh, H.; Reineke, U.; Schneider-Mergener, J.; Germeroth, L. "Spatially addressed synthesis of amino- and amino-oxy-substituted 1,3,5-triazine arrays on polymeric membranes." *J. Comb. Chem.*, **2000**, *2*, pp. 361-69.

41. Vanden Eynde, J.J.; Rutot, D. "Microwave-mediated derivatization of poly(styrene-*co*-allyl alcohol), a key step for the soluble polymer-assisted synthesis of heterocycles." *Tetrahedron* **1999**, *55*, pp. 2687-94.

42. Yu, A.M.; Zhang, Z.P.; Yang, H.Z.; Zhang, C.X.; Liu, Z. "Wang resin bound addition reactions under microwave irradiation." *Synth. Commun.* **1999**, *29*, 1595-99.

43. Gupta, M.; Paul, S.; Gupta, R. "Synthesis of 1,4-dithiocarbonyl piperazines under microwave irradiation in solvent-free conditions." *Synth. Commun.* **2001**, *31*, pp. 53-59.

44. Paul, S.; Gupta, M.; Gupta, R. "Vilsmeier reagent for formylation in solvent-free conditions using microwaves." *Synlett.* **2000**, *8*, pp. 1115-18.

45. Paul, S.; Gupta, M.; Gupta, R.; Loupy, A. "Microwave assisted synthesis of 1,5-disubstituted hydantoins and thiohydantoins in solvent-free conditions." *Synthesis* **2002**, *1*, pp. 75-78.

46. Paul, S.; Gupta, R.; Loupy, A.; Rani, B.; Dandia, A. "Dry media synthesis of 4H-1,4-benzothiazines under microwave irradiation using basic alumina as solid support." *Synth. Commun.* **2001**, *31*, pp. 711-17.

47. Paul, S.; Gupta, M.; Gupta, R.; Loupy, A. "Microwave assisted solvent-free synthesis of pyrazolo[3,4-b]quinolines and pyrazolo[3,4-c]pyrazoles using *p*-TsOH." *Tetrahedron Lett.* **2001**, *42*, pp. 3827-29.

48. Genta, M.T.; Villa, C.; Mariani, E.; Loupy, A.; Petit, A.; Rizzetto, R.; Mascarotti, A.; Morini, F.; Ferro, M. "Microwave-assisted preparation of cyclic ketals from a cineole ketone as potential cosmetic ingredients: solvent-free synthesis, odour evaluation, *in vitro* cytotoxicity, and antimicrobial assays." *Intl. J. Pharmaceutics* **2002**, *231*, pp. 11-20.

49. Rodriguez, H.; Perez, R.; Suarez, M.; Lam, A.; Cabrales, N.; Loupy, A. "Alkylation of some pyrimidine and purine derivatives in the absence of solvent using microwave-assisted method." *Fifth International Electronic Conference on Synthetic Organic Chemistry (ECSOC-5)* **2001**, E0004 (www.mdpi.net).

50. Villa, C.; Genta, M.T.; Bargagna, A.; Mariani, E.; Loupy, A. "Microwave activation and solvent-free phase transfer catalysis for the synthesis of new benzylidene cineole derivatives as potential UV sunscreens." *Green Chem.* **2001**, *3*, pp. 196-200.

51. Cleophax, J.; Liagre, M.; Loupy, A.; Petit, A. "Application of focused microwaves to the scale-up of solvent-free organic reactions." *Org. Process Res. Dev.* **2000**, *4*, pp. 498-504.

52. Loupy, A.; Régnier, S. " Solvent-free microwave-assisted Beckmann rearrangement of benzaldehyde and 2-hydroxyacetophenone oximes." *Tetrahedron Lett.* **1999**, *40*, pp. 6221-24.

53. Gelo-Pujic, M.; Guibe-Jampel, E.; Loupy, A.; Trincone, A. "Enzymatic glycosidation in dry media under microwave irradiation." *J. Chem. Soc., Perkin Trans. 1* **1997**, pp. 1001-02.

54. Limousin, C.; Cleophaz, J.; Petit, A.; Loupy, A.; Lukacs, G. "Solvent-free synthesis of decyl D-glycopyranosides: under focused microwave irradiation." *J. Carbohydrate Chem.* **1997**, *16*, pp. 327-42.

55. Sotelo, E.; Mocelo, R.; Suarez, M.; Loupy, A. "Synthesis of polyfunctional pyridazine derivatives using a solvent-free microwave assisted method." *Synth. Commun.* **1997**, *27*, pp. 2419-23.

56. Loupy, A.; Pigeon, P.; Ramdani, M. "Synthesis of long chain aromatic esters in a solvent-free procedure under microwaves." *Tetrahedron* **1996**, *52*, pp. 6705-12.

57. Marrero-Terrero, A.L.; Loupy, A. "Synthesis of 2-oxazolines from carboxylic acids and α,α,α-tris(hydroxymethyl)methylamine under microwaves in solvent-free conditions." *Synlett.* **1996**, *3*, pp. 245-46.

58. Perez, E.; Sotelo, E.; Loupy, A.; Mocelo, R.; Suarez, M.; Perez, R.; Autie, M. "An easy and efficient microwave-assisted method to obtain 1-(4-bromophenacyl)azoles in dry media." *Heterocycles* **1996**, *47*, pp. 539-43.

59. Suarez, M.; Loupy, A.; Perez, E.; Moran, L.; Gerona, G.; Morales, A.; Autie, M. "An efficient procedure to obtain hexahydroquinomeines and unsymmetrical 1,4-dihydropyridines using solid inorganic supports and microwave activation." *Heterocycl. Commun.* **1996**, *2*, pp. 275-80.

60. Bosch, A.; de la Cruz, P.; Diez-Barra, E.; Loupy, A.; Langa, F. "Microwave assisted Beckmann rearrangement of ketoximes in dry media." *Synlett.* **1995**, *12*, pp. 1259-60.

61. Loupy, A.; Petit, A.; Bonnet-Delpon, D. "Improvements in 1,3-dipolar cycloaddition of nitrones to flourinated dipolarophiles under solvent-free microwave activation." *J. Flourine Chem.* **1995**, *75*, pp. 215-16.

62. Diaz-Ortiz, A.; Prieto, P.; Loupy, A.; Abenhaim, D. "A short and efficient synthesis of ketene O,O- and S,S-acetals under focused microwave irradiation and solvent-free conditions." *Tetrahedron Lett.* **1996**, *37*, pp. 1695-98.

63. Diaz-Ortiz, A.; Diez-Barra, E.; de la Hoz, A.; Moreno, A.; Gomez-Escalonilla, M.J.; Loupy, A. "1,3-Dipolar cycloaddition of nitriles under microwave irradiation in solvent-free conditions." *Heterocycles* **1996**, *43*, pp. 1021-30.

64. Diaz-Ortiz, A.; de la Hoz, A.; Langa, F. "Microwave irradiation in solvent-free conditions: an eco-friendly methodology to prepare indazoles, pyrazolopyridines and bipyrazoles by cycloaddition reactions." *Green Chem.* **2000**, *2*, pp. 165-72.

65. Vaquez, E.; de la Hoz, A.; Elander, N.; Moreno, A.; Stone-Elander, S. "Microwave-assisted cyclocondensation under solvent-free conditions: quinoxaline-2,3-dione." *Heterocycles* **2001**, *55*, pp. 109-13.

66. de la Cruz, P.; de la Hoz, A.; Font, L.M.; Langa, F.; Pérez-Rodríguez, M.C. "Solvent-free phase transfer catalysis under microwaves in fullerene chemistry. A convenient preparation of N-alkylpyrrolidino[60]fullerenes." *Tetrahedron Lett.* **1998**, *39*, pp. 6053-56.

67. Balalaie, S.; Sharifi, A.; Ahangarian, B.; Kowsari, E. "Microwave enhanced synthesis of quinazolines in solvent-free condition." *Heterocycl. Commun.* **2001**, *7*, pp. 337-40.

68. Balalaie, S.; Kowsari, E. "One-pot synthesis of N-substituted 4-aryl-1,4-dihydropyridines under solvent-free condition and microwave irradiation." *Fifth International Electronic Conference on Synthetic Organic Chemistry (ECSOC-5)* **2001**, E0026 (www.mdpi.net).

69. Balalaie, S.; Salimi, S.H.; Sharifi, A. "Solid state deoximation with zinc chlorochromate: regeneration of carbonyl compounds." *Indian J. Chem., Sect. B* **2001**, *40*, pp. 1251-52.

70. Balalaie, S.; Golizeh, M. "Solvent-free organic synthesis on mineral supports using microwave irradiation." *Fifth International Electronic Conference on Synthetic Organic Chemistry (ECSOC-5)* **2001**, E0025 (www.mdpi.net).

71. Balalaie, S.; Hashtroudi, M.S.; Sharifi, A. "Microwave-assisted synthesis of 1,3,5-trialkyltetrahydro-1,3,5-triazin-2(1H)-ones and a 4-oxooxadiazinane in dry media." *J. Chem. Res. (S)* **1999**, pp. 392-93.

72. Ballini, R.; Bosica, G.; Fiorini, D. "Stereoselective preparation of (E)-ε-nitro-β,γ-unsaturated methyl esters: Amberlyst A 27, using microwave, as superior catalyst for the 1,6-conjugate addition of nitroalkanes to methyl 1,3-butadiene-1-carboxylate." *Tetrahedron Lett.* **2001**, *42*, pp. 8471-73.

73. Chemat, F.; Poux, M. "Microwave assisted pyrolysis of urea supported on graphite under solvent-free conditions." *Tetrahedron Lett.* **2001**, *42*, pp. 3693-95.

74. Danks, T.N.; Desai, B. "Microwave assisted chemistry using supported formates as reagents in organic chemistry." *Fifth International Electronic Conference on Synthetic Organic Chemistry (ECSOC-5)* **2001**, E0045 (www.mdpi.net).

75. Hajipour, A.R.; Mallakpour, S.E.; Imanzadeh, G. "An efficient and novel method for the synthesis of aromatic sulfones under solvent-free conditions." *Indian J. Chem. Sec. B* **2001**, *40*, pp. 237-39.

76. Hajipour, A.R.; Mallakpour, S.E.; Mohammadpoor-Baltork, I.; Khoee, S. "An efficient and selective method for conversion of oximes and semicarbazones to the corresponding carbonyl compounds under solvent-free conditions." *Synth. Commun.* **2001**, *31*, pp. 1187-94.

77. Hajipour, A.R.; Ghasemi, M. "A rapid and convenient synthesis of amides from aromatic acids and aliphatic amines in dry media under microwave irradiation." *Indian J. Chem. Sec. B* **2001**, *40*, pp. 504-07.

78. Hajipour, A.R.; Mallakpour, S.E.; Afrousheh, A. "A convenient and mild procedure for the synthesis of alkyl *p*-toluenesulfinates under solvent-free conditions using microwave irradiation." *Tetrahedron* **1999**, *55*, pp. 2311-16.

79. Heravi, M.M.; Tajbakhsh, M. "Solid state deoximation with clay supported potassium ferrate under microwave irradiation." *Phosphorus Sulfur Silicon Relat. Elem.* **2001**, *176*, pp. 195-99.

80. Heravi, M.M.; Ajami, D.; Mohajerani, B.; Tajbakhsh, M.; Ghassemzadeh, M.; Tabar-Hydar, K. "Solid state desemicarbazonation on clayfen under microwave irradiation." *Monatsh. Chem.* **2001**, *132*, pp. 881-83.

81. Heravi, M.M.; Rajabzadeh, G.; Rahimizadeh, M.; Bakavoli, M.; Ghassemzadeh, M. "Thiation of heterocycles using silica gel supported P_2S_5 under microwave irradiation in solventless system." *Synth. Commun.* **2001**, *31*, pp. 2231-34.

82. Tajbakhsh, M.; Heravi, M.M.; Habibzadeh, S.; Ghassemzadeh, M. "Microwave-assisted eco-friendly cleavage of acetals using supported potassium ferrate." *Phosphorus Sulfur Silicon Relat. Elem.* **2001**, *176*, pp. 151-55.

83. Tajbakhsh, M.; Heravi, M.M.; Habibzadeh, S. "Potassium ferrate supported on silica gel: a mild, efficient, and inexpensive reagent for oxidative deprotection of tetrahydropyranyl ethers in nonaqueous conditions." *Phosphorus Sulfur Silicon Relat. Elem.* **2001**, *176*, pp. 191-94.

84. Jeselnik, M.; Varma, R.S.; Polanc, S.; Kocevar, M. "Catalyst-free reactions under solvent-free conditions: microwave-assisted synthesis of heterocyclic hydrazones below the melting points of neat reactants." *J. Chem. Soc., Chem. Commun.* **2001**, pp. 1716-17.

85. Jeselnik, M.; Varma, R.S.; Polanc, S.; Kocevar, M. "Solid-state synthesis of heterocyclic hydrazones using microwaves under catalyst-free conditions." *Fifth International Electronic Conference on Synthetic Organic Chemistry (ECSOC-5)* **2001**, E0014 (www.mdpi.net).

86. Varma, R.S. "Solvent-free accelerated organic syntheses using microwaves." *Pure Appl. Chem.* **2001**, *73*, pp. 193-98.

87. Varma, R.S. "Expeditious solvent-free organic syntheses using microwave irradiation." *Green Chem. Synth. Proc.* **2000**, *767*, pp. 292-312.

88. Varma, R.S.; Kumar, D. "Microwave-accelerated solvent-free synthesis of thioketones, thiolactones, thioamides, thionoesters, and thioflavonoids." *Org. Lett.* **1999**, *1*, pp. 697-700.

89. Varma, R.S. "Solvent-free organic synthesis using supported reagents and microwave irradiation." *Green Chem.* **1999**, *1*, pp. 43-55.

90. Varma, R.S.; Naicker, K.P. "Solvent-free synthesis of amides from non-enolizable esters and amines using microwave irradiation." *Tetrahedron Lett.* **1999**, *40*, pp. 6177-80.

91. Varma, R.S. "Solvent-free synthesis of heterocyclic compounds using microwaves." *J. Heterocycl. Chem.* **1999**, *36*, pp. 1565-71.

92. Vass, A.; Dudas, J.; Varma, R.S. "Solvent-free synthesis of N-sulfonylimines using microwave irradiation." *Tetrahedron Lett.* **1999**, *40*, pp. 4951-54.

93. Varma, R.S.; Naicker, K.P. "Hydroxylamine on clay: a direct synthesis of nitriles from aromatic aldehydes using microwaves under solvent-free conditions." *Molecules Online* **1998**, *2*, pp. 94-96.

94. Varma, R.S.; Naicker, K.P.; Liesen, P.J. "Microwave-accelerated crossed Cannizzaro reaction using barium hydroxide under solvent-free conditions." *Tetrahedron Lett.* **1998**, *39*, pp. 8437-40.

95. Varma, R.S.; Naicker, K.P.; Liesen, P.J. "Selective nitration of styrenes with clayfen and clayan: a solvent-free synthesis of β-nitrostyrenes." *Tetrahedron Lett.* **1998**, *39*, pp. 3977-80.

96. Varma, R.S.; Saini, R.K. "Microwave-assisted isomerization of 2'-aminochalcones on clay: an easy route to 2-aryl-1,2,3,4-tetrahydro-4-quinolones." *Synlett.* **1997**, *87*, pp. 857-58.

97. Varma, R.S.; Dahiya, R.; Kumar, S. "Clay catalyzed synthesis of imines and enamines under solvent-free conditions using microwave irradiation." *Tetrahedron Lett.* **1997**, *38*, pp. 2039-42.

98. Varma, R.S.; Dahiya, R. "An expeditious and solvent-free synthesis of 2-amino-substituted isoflav-3-enes using microwave irradiation." *J. Org. Chem.* **1998**, *63*, pp. 8038-41.

99. Kabalka, G.W.; Pagni, R.M.; Wang, L.; Namboodiri, V. "Microwave-assisted, solventless Suzuki coupling reactions on palladium-doped alumina." *Green Chem.* **2000**, *2*, pp. 120-22 and *Fifth International Electronic Conference on Synthetic Organic Chemistry (ECSOC-5)* **2001**, E0029 (www.mdpi.net).

100. Kabalka, G.W.; Wang, L.; Namboodiri, V.; Pagni, R.M. "Rapid microwave-enhanced, solventless Sonogashira coupling reaction on alumina." *Tetrahedron Lett.* **2000**, *41*, pp. 5151-54.

101. Kabalka, G.W.; Wang, L.; Pagni, R.M. "Microwave-enhanced Glaser coupling under solvent free conditions." *Synlett.* **2001**, *1*, pp. 108-10.

102. Kaboudin, B.; Nazari, R. "A convenient and mild procedure for the preparation of α-ketophosphonates of 1-hydroxyphosphonates under solvent-free conditions using microwave." *Synth. Commun.* **2001**, *31*, pp. 2245-50.

103. Kaboudin, B.; Nazari, R. "Microwave-assisted synthesis of 1-aminoalkylphosphonates under solvent-free conditions." *Tetrahedron Lett.* **2001**, *42*, pp. 8211-13.

104. Kidwai, M.; Misra, P.; Bhushan, K.R. "Alumina-supported synthesis of thiadiazolyl thiazolothiones." *Synth. Commun.* **2001**, *31*, pp. 817-22.

105. Kidwai, M. "Dry media reactions." *Pure Appl. Chem.* **2001**, *73*, pp. 147-51 and *Fifth International Electronic Conference on Synthetic Organic Chemistry (ECSOC-5)* **2001**, E0032 (www.mdpi.net).

106. Kidwai, M.; Sapra, P. "An expeditious solventless synthesis of isoxazoles." *Org. Prep. Proced. Intl.* **2001**, *33*, pp. 381-86.

107. Kidwai, M.; Sapra, P.; Misra, P.; Saxena, R.K.; Singh, M. "Microwave-assisted solid support synthesis of novel 1,2,4-triazolo[3,4-b]-1,3,4-thiadiazepines as potent antimicrobial agents." *Bioorg. Med. Chem.* **2001**, *9*, pp. 217-220.

108. Kidwai, M.; Sapra, P.; Bhushan, K.R.; Misra, P. "Microwave-assisted solid support synthesis of pyrazolino/iminopyrimidino/thioxopyrimidino imidazolines." *Synthesis* **2001**, *10*, pp. 1509-12.

109. Kidwai, M.; Sapra, P.; Bhushan, K.R.; Misra, P. "Microwave-assisted synthesis of novel 1,2,4-triazines in 'dry media'." *Synth. Commun.* **2001**, *31*, pp. 1639-45.

110. Kidwai, M., Bhushan, K.R.; Sapra, P.; Saxena, R.K.; Gupta, R. "Alumina-supported synthesis of antibacterial quinolines using microwaves." *Bioorg. Med. Chem.* **2000**, *8*, pp. 69-72.

111. Kidwai, M.; Venkataramanan, R.; Kohli, S. "Alumina-supported synthesis of β-lactams using microwave." *Synth. Commun.* **2000**, *30*, pp. 989-1002.

112. Kidwai, M.; Misra, P.; Dave, B.; Bhushan, K.R.; Saxena, R.K.; Singh, M. "Microwave-activated solid support synthesis of new antibacterial quinolones." *Monatsh. Chem.* **2000**, *131*, pp. 1207-12.

113. Kidwai, M.; Misra, P.; Bhushan, K.R. "Microwave assisted synthesis of novel organomercurials in 'dry media'." *Polyhedron* **1999**, *18*, pp. 2641-43.

114. Lange, J.H.M.; Verveer, P.C.; Osnabrug, S.J.M.; Visser, G.M. "Rapid microwave-enhanced synthesis of 4-hydroxyquinolinones under solvent-free conditions." *Tetrahedron Lett.* **2001**, *42*, pp. 1367-69.

115. Li, J.P.; Luo, Q.F.; Song, Y.M.; Wang, Y.L. "A rapid and efficient synthesis of diaryl thioureas via solvent-free reaction using microwave." *Chinese Chem. Lett.* **2001**, *12*, pp. 383-86.

116. Li, J.P.; Luo, Q.F.; Wang, Y.L.; Wang, H. "Solvent-free synthesis of heterocyclic thioureas using microwave technology." *J. Chin. Chem. Soc.* **2001**, *48*, pp. 73-75.

117. Wang, C.; Li, G.S.; Li, J.C.; Feng, S.; Li, X.L. "Synthesis of 5-alkylene barbituric acid in solventless system under microwave irradiation." *Chin. J. Org. Chem.* **2001**, *21*, pp. 310-12.

118. Xu, Q.; Chao, B.; Wang, Y.; Dittmer, D.C. "Tellurium in the "no-solvent" organic synthesis of allylic alcohols." *Tetrahedron* **1997**, *53*, pp. 12131-146.

119. Loghmani-Khouzani, H.; Sadeghi, M.M.; Safari, J.; Minaeifar, A. "A novel method for the synthesis of 2-ketomethylquinolines under solvent-free conditions using microwave irradiation." *Tetrahedron Lett.* **2001**, *42*, pp. 4363-64.

120. Loghmani-Khouzani, H.; Sadeghi, M.M.; Safari, J.; Abdorrezaie, M.S.; Jafarpisheh, M. "A convenient synthesis of azines under solvent-free conditions using microwave irradiation." *J. Chem. Res. (S)* **2001**, pp. 80-81.

121. Shaabani, A.; Bahadoran, F.; Bazgir, A.; Safari, N. "Synthesis of metallophthalocyanines under solvent-free conditions using microwave irradiation." *Iranian J. Chem. & Chem. Eng. (Intl. Engl. Ed.)* **1999**, *18*, pp. 104-07.

122. Maree, M.D.; Nyokong, T. "Solvent-free axial ligand substitution in octaphenoxyphthalocyaninato silicon complexes using microwave irradiation." *J. Chem. Res. (S)* **2001**, pp. 68-69.

123. Massicot, F.; Plantier-Royon, R.; Portella, C.; Saleur, D.; Sudha, A.V.R.L. "Solvent-free synthesis of tartramides under microwave activation." *Synthesis* **2001**, pp. 2441-44.

124. Mirjalili, B.F.; Zolfigol, M.A.; Bamoniri, A. "Silica sulfuric acid/wet SiO_2 as a novel system for the deprotection of acetals by using microwave irradiation under solvent free conditions." *J. Korean Chem. Soc.* **2001**, *45*, pp. 546-48.

125. Pezet, F.; Sasaki, I.; Daran, J.C.; Hydrio, J.; Ait-Haddou, H.; Balavoine, G. "First example of supported microwave-assisted synthesis of new chiral bipyridines and a terpyridine - use in asymmetric cyclopropanation." *Eur. J. Inorg. Chem.* **2001**, pp. 2669-74.

126. Quiroga, J.; Cisneros, C.; Insuasty, B.; Abonia, R.; Nogueras, M.; Sanchez, A. "A regiospecific three-component one-step cyclocondensation to 6-cyano-5,8-dihydropyrido-[2,3-d]pyrimidin-4(3H)-ones using microwaves under solvent-free conditions." *Tetrahedron Lett.* **2001**, *42*, pp. 5625-27.

127. Rajakumar, P.; Murali, V. "Sodium sulfate supported synthesis of cationic cyclophanes using microwaves." *J. Chem. Soc., Chem. Commun.* **2001**, pp. 2710-11.

128. Romanova, N.N.; Kudan, P.V.; Gravis, A.G.; Zyk, N.V. "Investigation of the stereochemistry of rapid solvent-free microwave syntheses of β-amino acid esters." *Fifth International Electronic Conference on Synthetic Organic Chemistry (ECSOC-5)* **2001**, E0018 (www.mdpi.net).

129. Romanova, N.N.; Gravis, A.G.; Kudan, P.V.; Bundel, Y.G. "Solvent-free stereoselective synthesis of β-aryl-β-amino acid esters by the Rodionov reaction using microwave irradiation." *Mendeleev Commun.* **2001**, pp. 26-27.

130. Shanmugam, P.; Singh, P.R. "Montmorillonite K-10 clay-microwave assisted isomerisation of acetates of the Baylis-Hillman adducts: a facile method of stereoselective synthesis of (E)-trisubstituted alkenes." *Synlett.* **2001**, *8*, pp. 1314-16.

131. Vanden Eynde, J.J.; Hecq, N.; Kappe, C.O.; Kataeva, O. "Microwave-mediated regioselective synthesis of novel pyrimido[1,2-a]pyrimidines under solvent-free conditions." *Tetrahedron* **2001**, *57*, pp. 1785-91 and *Fourth International Electronic Conference on Synthetic Organic Chemistry (ECSOC-4)* **2000**, A0050 (www.mdpi.net).

132. a) Villemin, D.; Hammadi, M.; Hachemi, M.; Bar, N. "Applications of microwave in organic synthesis: an improved one-step synthesis of metallophthalocyanines and a new modified microwave oven for dry reaction." *Fifth International Electronic Conference on Synthetic Organic Chemistry (ECSOC-5)* **2001**, E0046 (www.mdpi.net). b) Villemin, D.; Hammadi, M.; Hachemi, M. *Synth. Commun.* **2002**, *32*, in press.

133. Villemin, D.; Hammadi, M.; Martin, B. "Clay catalysis: condensation of orthoesters with *o*-substituted aminoaromatics into heterocycles." *Synth. Commun.* **1996**, *26*, pp. 2895-99.

134. Villemin, D.; Hammadi, M. "Environmentally desirable synthesis without the use of organic solvent. Synthesis of aryloxyacetic acids." *Synth. Commun.* **1996**, *26*, pp. 4337-41.

135. Villemin, D.; Hachemi, M.; Lalaoui, M. "Potassium fluoride on alumina: synthesis of O-aryl-N,N-dimethylthiocarbamates and their rearrangement into S-aryl-N,N-dimethylthiocarbamates under microwave irradiation." *Synth. Commun.* **1996**, *26*, pp. 2461-71.

136. Villemin, D.; Martin, B. "Dry condensation of creatinine with aldehydes under focused microwave irradiation." *Synth. Commun.* **1995**, *25*, pp. 3135-40.

137. Villemin, D.; Hammadi, M. "Oxidation by DMSO. II. Opening of epoxides into a-hydroxyketones in presence of KSF clay under microwave irradiation." *Synth. Commun.* **1995**, *25*, pp. 3141-44.

138. Yadav, J.S.; Reddy, B.V.S.; Madan, C. "Microwave-assisted efficient one-pot synthesis of nitriles in dry media." *J. Chem. Res. (S)* **2001**, pp. 190-91.

139. Yadav, J.S.; Reddy, B.V.S.; Madan, C. "Montmorillonite clay-catalyzed one-pot synthesis of α-amino phosphonates." *Synlett.* **2001**, *7*, pp. 1131-33.

140. Yadav, J.S.; Reddy, B.V.S.; Madan, C. "Montmorillonite clay-catalyzed stereoselective syntheses of aryl-substituted (E)- and (Z)-allyl iodides and bromides." *New J. Chem.* **2001**, *25*, pp. 1114-17.

141. Meshram, H.M.; Sekhar, K.C.; Ganesh, Y.S.S.; Yadav, J.S. "Clay catalyzed facile cyclodehydration under microwave: synthesis of 3-substituted benzofurans." *Synlett.* **2000**, *9*, pp. 1273-74.

142. Ramalingam, T.; Reddy, B.V.S.; Srinivas, R.; Yadav, J.S. "Solvent-free conversion of N,N-dimethylhydrazones to nitriles under microwave irradiation." *Synth. Commun.* **2000**, *30*, pp. 4507-12.

143. Kumar, H.M.S.; Reddy, B.V.S.; Anjaneyulu, S.; Reddy, E.J.; Yadav, J.S. "Clay catalysed amidation of alcohols with nitriles in dry media." *New J. Chem.* **1999**, pp. 955-56.

144. Sabitha, G.; Babu, R.S.; Yadav, J.S. "One pot synthesis of 4-(2-hydroxybenzoyl) pyrazoles from 3-formylchromones under microwave irradiation in solvent free conditions." *Synth. Commun.* **1998**, *28*, pp. 4571-76.

145. Bandgar, B.P.; Kasture, S.P. "Microwave-induced solvent-free, rapid and efficient synthesis of conjugated nitroalkenes using sulfated zirconia." *Indian J. Chem., Sect. B* **2001**, *40*, pp. 1239-41.

146. Bandgar, B.P.; Makone, S.S. "Rapid and selective regeneration of carbonyl compounds from their oximes under mild, neutral and solvent-free conditions." *Org. Prep. Proced. Intl.* **2000**, *32*, pp. 391-94.

147. Bandgar, B.P.; Uppalla, L.S.; Kurule, D.S. "Solvent-free one-pot rapid synthesis of 3-carboxycoumarins - using focused microwaves." *Green Chem.* **1999**, *1*, pp. 243-45.

148. Michaud, D.; Abdallah-El Ayoubi, S.; Dozias, M.J.; Toupet, L.; Texier-Boullet, F.; Hamelin, J. "New route to functionalized cyclohexenes from nitromethane and electrophilic alkenes without solvent under focused microwave irradiation." *J. Chem. Soc., Chem. Commun.* **1997**, pp. 1613-14.

149. Kasmi, S.; Hamelin, J.; Benhaoua, H. "Microwave-assisted solvent-free synthesis of iminothiazolines." *Tetrahedron Lett.* **1998**, *39*, pp. 8093-96.

150. Perio, B.; Dozias, M.J.; Hamelin, J. "Ecofriendly fast batch synthesis of dioxolanes, dithiolanes, and oxathiolanes without solvent under microwave irradiation." *Org. Process Res. Dev.* **1998**, *2*, pp. 428-30.

151. Saoudi, A.; Hamelin, J.; Benhaoua, H. "Elimination reaction over solid supports under microwave irradiation: synthesis of functionalized alkenes." *Tetrahedron Lett.* **1998**, *39*, pp. 4035-38.

152. Meddad, N.; Rahmouni, M.; Derdour, A.; Bazureau, J.P.; Hamelin, J. "Eco-friendly transamination and aza-annulation reactions: solvent-free synthesis of new α-hetero-β-hydrazino acrylates and 1,2-dihydropyrazol-3-ones." *Synthesis* **2001**, pp. 581-84.

153. Benhaliliba, H.; Derdour, A.; Bazureau, J.P.; Texier-Boullet, F.; Hamelin, J. "Solvent free oxidation of β,β-disubstituted enamines under microwave irradiation." *Tetrahedron Lett.* **1998**, *39*, pp. 541-42.

154. Dahmani, Z.; Rahmouni, M.; Brugidou, R.; Bazureau, J.P.; Hamelin, J. "A new route to α-hetero-β-enamino esters using a mild and convenient solvent-free process assisted by focused microwave irradiation." *Tetrahedron Lett.* **1998**, *39*, pp. 8453-56.

155. Kerneur, G.; Lerestif, J.M.; Bazureau, J.P.; Hamelin, J. "Convenient preparation of 4-alkylidene-lH-imidazol-5(4H)-one derivatives from imidate and aldehydes by a solvent-free cycloaddition under microwaves." *Synthesis* **1997**, *3*, pp. 287-89.

156. Cherouvrier, J.R.; Bazureau, J.P. "A practical and stereoselective route to 5-ylidene-3,5-dihydroimidazol-4-one derivatives using solvent-free conditions under focused microwave irradiations." *Fifth International Electronic Conference on Synthetic Organic Chemistry (ECSOC-5)* **2001**, E0012 (www.mdpi.net).

157. Fraga-Dubreuil, J.; Cherouvrier, J.R.; Bazureau, J.P. "Clean solvent-free dipolar cycloaddition reactions assisted by focused microwave irradiations for the synthesis of new ethyl 4-cyano-2-oxazoline-4-carboxylates." *Green Chem.* **2000**, *2*, pp. 226-29.

158. Gianotti, M.; Martelli, G.; Spunta, G.; Campana, E.; Panunzio, M.; Mendozza, M. "Solvent-free microwave-assisted organic reactions, preparation of β-keto-esters." *Synth. Commun.* **2000**, *30*, pp. 1725-30.

159. Esteves-Souza, A.; Echevarria, A.; Vencato, I.; Jimeno, M.L.; Elguero, J. "Unexpected formation of bis-pyrazolyl derivatives by solid support coupled with microwave irradiation." *Tetrahedron* **2001**, *57*, pp. 6147-53.

160. Gomez-Lara, J.; Gutierrez-Perez, R.; Penieres-Carrillo, G.; Lopez-Cortes, J.G.; Escudero-Salas, A.; Alvarez-Toledano, C. "Reaction of hydroquinones with supported oxidizing reagents in solvent-free conditions." *Synth. Commun.* **2000**, *30*, pp. 2713-20.

161. Lopez-Cortes, J.G.; Penieres-Carrillo, G.; Ortega-Alfaro, M.C.; Gutierrez-Perez, R.; Toscano, R.A.; Alvarez-Toledano, C. "Oxidative coupling-type mechanism of N,N-dialkylanilines in solvent-free conditions forming crystal violet derivatives. A clay-mediated and microwave-promoted approach." *Can. J. Chem.* **2000**, *78*, pp. 1299-304.

162. Rico-Gomez, R.; Najera, F.; Lopez-Romero, J.M.; Canada-Rudner, P. "Solvent-free synthesis of thio-alkylxanthines from alkylxanthines using microwave irradiation." *Heterocycles* **2000**, *53*, pp. 2275-78.

163. Ley, S.V.; Baxendale, I.R.; Bream, R.N.; Jackson, P.S.; Leach, A.G.; Longbottom, D.A.; Nesi, M.; Scott, J.S.; Storer, R.I.; Taylor, S.J. "Multi-step organic synthesis using solid-supported reagents and scavengers: a new paradigm in chemical library generation." *J. Chem. Soc., Perkin Trans. 1* **2000**, pp. 3815-4195.

164. Habermann, J.; Ley, S.V.; Scott, J.S. "Synthesis of the potent analgesic compound (±)-epibatidine using an orchestrated multi-step sequence of polymer supported reagents." *J. Chem. Soc., Perkin Trans. 1* **1999**, pp. 1253-56.

165. Olsson, R.; Hansen, H.C.; Andersson, C.M. "Microwave-assisted solvent-free parallel synthesis of thioamides." *Tetrahedron Lett.* **2000**, *41*, pp. 7947-50.

166. Sauvagnat, B.; Lamaty, F.; Lazaro, R.; Martinez, J. "Poly(ethyleneglycol) as solvent and polymer support in the microwave assisted parallel synthesis of amino acid derivatives." *Tetrahedron Lett.* **2000**, *41*, pp. 6371-75.

167. Varray, S.; Sauvagnat, B.; Gauzy, C.; Lamaty, F.; Lazaro, R.; Martinez, J. "Poly(ethyleneglycol) supported synthesis of amino acid derivatives via ring closing metathesis or microwave-assisted alkylation." *Fourth International Electronic Conference on Synthetic Organic Chemistry (ECSOC-4)* **2000**, B0010 (www.mdpi.net).

168. Sharifi, A.; Mohsenzadeh, F.; Naimi-Jamal, M.R. "Solvent-free preparation of monoacylaminals assisted by microwave irradiation." *J. Chem. Res. (S)* **2000**, pp. 394-96.

169. Stefani, H.A.; Gatti, P.M. "3,4-Dihydropyrimidin-2(1H)-ones: fast synthesis under microwave irradiation in solvent-free conditions." *Synth. Commun.* **2000**, *30*, pp. 2165-73.

170. Tamami, B.; Kiasat, A.R. "Microwave promoted rapid oxidative deoximation of oximes under solvent-free conditions." *Synth. Commun.* **2000**, *30*, pp. 4129-35.

171. Jnaneshwara, G.K.; Deshpande, V.H.; Bedekar, A.V. "Clay-catalyzed conversion of 2,2-disubstituted malononitriles to 2-oxazolines: towards unnatural amino acids." *J. Chem. Res. (S)* **1999**, pp. 252-53.

172. Mojtahedi, M.M.; Saidi, M.R.; Bolourtchian, M. "Microwave assisted aminolysis of epoxides under solvent-free conditions catalyzed by montmorillonite clay." *J. Chem. Res. (S)* **1999**, pp. 128-29.

173. Bogdal, D. "Coumarins - solvent free synthesis by the Knoevenagel condensation under microwave irradiation." *Electronic Conference on Trends in Heterocyclic Chemistry (ECTOC-4: ECHET98)* **1998**, Article 087 (www.ch.ic.ac.uk/ectoc/).

174. Bogdal, D.; Pielichowski, J.; Jaskot, K. "A rapid Williamson synthesis under microwave irradiation in dry media." *Org. Prep. Proc. Intl.* **1998**, *30*, pp. 427-32.

175. Bogdal, D.; Pielichowski, J.; Boron, A. "Synthesis of aromatic ethers under microwave irradiation in dry media." *First International Electronic Conference on Synthetic Organic Chemistry (ECSOC-1)* **1997**, A0049 (www.mdpi.net).

176. Coville, N.J.; Cheng, L. "Organometallic chemistry in the solid state." *J. Organomet. Chem.* **1998**, *571*, pp. 149-69.

177. Corsaro, A.; Chiacchio, U.; Librando, V.; Fisichella, S.; Pistara, V. "1,3-Dipolar cycloadditions of polycyclic aromatic hydrocarbons with nitrile oxides under microwave irradiation in the absence of solvent." *Heterocycles* **1997**, *45*, pp. 1567-72.

178. Filip, S.V.; Nagy, G.; Surducan, E.; Surducan, V. "Microwave application in organic synthesis. Microwave-assisted preparation of diphenylamines in 'dry media'." *First International Electronic Conference on Synthetic Organic Chemistry (ECSOC-1)* **1997**, A0031 (www.mdpi.net).

179. Nagy, G.; Filip, S.V.; Surducan, E.; Surducan, V. "Solvent-free synthesis of substituted phenoxyacetic acids under microwave irradiation." *Synth. Commun.* **1997**, *27*, pp. 3729-36.

180. Kad, G.L.; Bhandari, M.; Kaur, J.; Rathee, R.; Singh, J. "Solventless preparation of oximes in the solid state and via microwave irradiation." *Green Chem.* **2001**, *3*, pp. 275-77.

181. Kad, G.L.; Singh, V.; Kaur, K.P.; Singh, J. "Selective preparation of benzylic bromides in dry media coupled with microwave irradiation." *Tetrahedron Lett.* **1997**, *38*, pp. 1079-80.

182. Majetich, G.; Wheless, K. *Microwave-Enhanced Chemistry Fundamentals, Sample Preparation, and Applications*, Kingston, H.M.; Haswell, S.J., Eds., American Chemical Society **1997**, ch. 8, pp. 455-505.

183. Lidstrom, P.; Tierney, J.; Wathey, B.; Westman, J. "Microwave assisted organic synthesis – a review." *Tetrahedron* **2001**, *57*, pp. 9225-83.

184. Mills, K.; United States Patent No. 4933461, "Preparation of a piperidinylcyclopentylheptenoic acid derivative." **1990**.

185. Collington, E.W.; Finch, H.; Hayes, R.; Mills, K.; Woodings, D.F.; United States Patent No. 5039673, "Aminocyclopentyl ethers and their preparation and pharmaceutical formulation." **1991**.

186. a) Wali, A.; Pillai, S.M.; Satish, S. "Heterogeneous Pd catalysts and microwave irradiation in Heck arylation." *Research Centre, Indian Petrochemicals Corporation Ltd.* **1995**, IPCL Communication No. 294. b) Wali, A.; Pillai, S.M.; Satish, S. "Heterogeneous Pd catalysts and microwave irradiation in Heck arylation." *React. Kinet. Catal. Lett.* **1997**, *60*, pp. 189-94 [Received on October 17, **1995**].

187. Villemin, D.; Jaffres, P.A.; Nechab B., ENSI de Caen, ISMRA, Caen, France, presented at the University of Saragosse, Spain, **1995**.

188. Larhed, M.; Hallberg, A. "Microwave-promoted palladium-catalyzed coupling reactions." *J. Org. Chem.* **1996**, *61*, pp. 9582-84.

189. Larhed, M.; Lindeberg, G.; Hallberg, A. "Rapid microwave-assisted Suzuki coupling on solid-phase." *Tetrahedron Lett.* **1996**, *37*, pp. 8219-22.

190. Larhed, M.; Hoshino, M.; Hadida, S.; Curran, D.P.; Hallberg, A. "Rapid fluourous Stille coupling reactions conducted under microwave irradiation." *J. Org. Chem.* **1997**, *62*, pp. 5583-87.

191. Bremberg, U.; Larhed, M.; Moberg, C.; Hallberg, A. "Rapid microwave-induced palladium-catalyzed asymmetric allylic alkylation." *J. Org. Chem.* **1999**, *64*, pp. 1082-83.

192. Olofsson, K.; Kim, S.Y.; Larhed, M.; Curran, D.P.; Hallberg, A. "High-speed, highly fluorous organic reactions." *J. Org. Chem.* **1999**, *64*, pp. 4539-41.

193. Kaiser, N.F.K.; Bremberg, U.; Larhed, M.; Moberg, C.; Hallberg, A. "Microwave-mediated palladium-catalyzed asymmetric allylic alkylation; an example of highly selective fast chemistry." *J. Organomet. Chem.* **2000**, *603*, pp. 2-5.

194. Bremberg, U.; Lutsenko, S.; Kaiser, N.F.; Larhed, M.; Hallberg, A.; Mobert, C. "Rapid and stereoselective C-C, C-O, C-N and C-S couplings via microwave accelerated palladium-catalyzed allylic substitutions." *Synthesis* **2000**, *7*, pp. 1004-08.

195. Kaiser, N.F.K.; Bremberg, U.; Larhed, M.; Moberg, C.; Hallberg, A. "Fast, convenient, and efficient molybdenum-catalyzed asymmetric allylic alkylation under non-inert conditions: an example of microwave-promoted fast chemistry." *Angew. Chem., Int. Ed. Eng.* **2000**, *39*, pp. 3596-98.

196. Vallin, K.S.A.; Larhed, M.; Johansson, K.; Hallberg, A. "Highly selective palladium-catalyzed synthesis of protected α,β-unsaturated methyl ketones and 2-alkoxy-1,3-butadienes. High-speed chemistry by microwave flash heating." *J. Org. Chem.* **2000**, *65*, pp. 4537-42.

197. Olofsson, K.; Sahlin, H.; Larhed, M.; Hallberg, A. "Regioselective palladium-catalyzed synthesis of β-arylated primary allylamine equivalents by an efficient Pd-N coordination." *J. Org. Chem.* **2001**, *66*, pp. 544-49.

198. Vallin, K.S.A.; Larhed, M.; Hallberg, A. "Aqueous DMF-potassium carbonate as a substitute for thallium and silver additives in the palladium-catalyzed conversion of aryl bromides to acetyl arenes." *J. Org. Chem.* **2001**, *66*, pp. 4340-43.

199. Garg, N.; Larhed, M.; Hallberg, A. "Heck arylation of 1,2-cyclohexanedione and 2-ethoxy-2-cyclohexenone." *J. Org. Chem.* **1998**, *63*, pp. 4158-62.

200. Olofsson, K.; Larhed, M.; Hallberg, A. "Highly regioselective palladium-catalyzed internal arylation of allyltrimethylsilane with aryl triflates." *J. Org. Chem.* **1998**, *63*, pp. 5076-79.

201. Moberg, C.; Bremberg, U.; Hallman, K.; Svensson, M.; Norrby, P.O.; Hallberg, A.; Larhed, M.; Csoregh, I. "Selectivity and reactivity in asymmetric allylic alkylation." *Pure Appl. Chem.* **1999**, *71*, pp. 1477-83.

202. Qabaja, G.; Jones, G.B. "An intramolecular arylation route to the kinafluorenones." *Tetrahedron Lett.* **2000**, *41*, pp. 5317-20.

203. Diaz-Ortiz, A.; Prieto, P.; Vazquez, E. "Heck reactions under microwave irradiation in solvent-free conditions." *Synlett.* **1997**, pp. 269-70.

204. Combs, A.P.; Saubern, S.; Rafalski, M.; Lam, P.Y.S. "Solid supported aryl/heteroaryl C-N cross-coupling reactions." *Tetrahedron Lett.* **1999**, *40*, pp. 1623-26.

205. Blettner, C.G.; Konig, W.A.; Stenzel, W.; Schotten, T. "Microwave-assisted aqueous Suzuki cross-coupling reactions." *J. Org. Chem.* **1999**, *64*, pp. 3885-90.

206. Villemin, D.; Gomez-Escalonilla, M.J.; Saint-Clair, J.F. "Palladium-catalysed phenylation of heteroaromatics in water or methylformamide under microwave irradiation." *Tetrahedron Lett.* **2001**, *42*, pp. 635-37.

207. Villemin, D.; Caillot, F. "Microwave mediated palladium-catalysed reactions on potassium fluoride/alumina without use of solvent." *Tetrahedron Lett.* **2001**, *42*, pp. 639-42.

208. Ohberg, L.; Westman, J. "One-pot three-step solution phase syntheses of thiohydantoins using microwave heating." *Synlett.* **2001**, *12*, pp. 1893-96.

209. Sasaki, S.; Ishibashi, N.; Kuwamura, T.; Sano, H.; Matoba, M.; Nisikawa, T.; Maeda, M. "Excellent acceleration of the Diels-Alder reaction by microwave irradiation for the synthesis of new fluorine-substituted ligands of NMDA receptor." *Bioorg. Med. Chem. Lett.* **1998**, *8*, pp. 2983-86.

210. Kamath, C.R.; Samant, S.D. "The Diels-Alder reaction of 2H-pyran-2-ones. Part IV. Microwave-catalyzed Diels-Alder reaction of 4,6-disubstituted-2H-pyran-2-ones with 1,4-naphthoquinone and N-phenylmaleimide." *Indian J. Chem.* **1996**, *35B*, pp. 256-59.

211. Zhu, R.; Hong, P.; Dai, S. "Study on the 'dry reaction' without any medium under microwave irradiation." *Synth. Commun.* **1994**, *24*, pp. 2417-21.

212. Sridhar, M.; Krishna, K.L.; Srinivas, K.; Rao, J.M. "Microwave-activated Diels-Alder cycloaddition reactions of 2-chloro-1,2-difluorovinyl phenyl sulfone." *Tetrahedron Lett.* **1998**, *39*, pp. 6529-32.

213. Mayoral, J.A.; Cativiela, C.; Garcia, J.I.; Pires, E.; Royo, A.J.; Figueras, F. "Diels-Alder reaction of α-amino acid precursors by heterogeneous catalysis: thermal vs. microwave activation." *Appl. Catal. A: General* **1995**, *131*, pp. 159-66.

214. Diaz-Ortiz, A. Carrillo, J.R.; Diez-Barra, E.; de la Hoz, A.; Gomez-Escalonilla, M.J.; Moreno, A.; Langa, F. "Diels-Alder cycloaddition of vinylpyrazoles. Synergy between microwave irradiation and solvent-free conditions." *Tetrahedron* **1996**, *52*, pp. 9237-48.

215. Yin, D.H.; Yin, D.L.; Li, Q.H. "Highly regioselective synthesis of *para*-myrac aldehyde catalyzed by modified HY zeolite under microwave irradiation." *Chin. Chem. Lett.* **1996**, *7*, pp. 697-98.

216. Lu, Y.F.; Fallis, A.G. "A cycloaddition approach to tricyclic taxoid skeletons." *Can. J. Chem.* **1995**, *73*, pp. 2239-52.

217. Takatori, K.; Hasegawa, K.; Narai, S.; Kajiwara, M. "A microwave-accelerated intramolecular Diels-Alder reaction approach to compactin." *Heterocycles* **1996**, *42*, pp. 525-28.

218. Jankowski, C.K.; LeClair, G.; Belanger, J.M.R.; Pare, J.M.J.; Van Calsteren, M.R. "Microwave-assisted Diels-Alder synthesis." *Can. J. Chem.* **2001**, *79*, pp. 1906-09.

219. Silveira, C.C.; Nunes, M.R.S.; Wendling, E.; Braga, A.L. "Synthesis of α-phenylseleno-α,β-unsaturated esters by Wittig-type reactions. Studies on the Diels-Alder reaction." *J. Organomet. Chem.* **2001**, *623*, pp. 131-36.

220. Yoon, H.J.; Lee, J.M.; Im, S.S.; Chae, W.K. "Triplex, microwave induced and conventional Diels-Alder reaction of phenyl-substituted alkynes with cyclopentadienes." *Bull. Korean Chem. Soc.* **2001**, *22*, pp. 543-44.

221. de la Hoz, A.; Diaz-Ortiz, A.; Moreno, A.; Langa, F. "Cycloadditions under microwave irradiation conditions: methods and applications." *Eur. J. Org. Chem.* **2000**, pp. 3659-73.

222. Sridhar, M.; Krishna, K.L.; Rao, J.M. "Synthesis and Diels–Alder cycloaddition reactions of [(2,2-dichloro-1-fluoroethenyl)sulfinyl]benzene and [(2-chloro-1,2-difluoro ethenyl)sulfinyl]benzene." *Tetrahedron* **2000**, *56*, pp. 3539-45.

223. Majetich, G.; Hicks, R. "The use of microwave heating to promote organic reactions." *J. Microwave Power Electromagnetic Energy* **1995**, *30*, pp. 27-45.

224. Avalos, M.; Babiano, R.; Cintas, P.; Clemente, F.R.; Jimenez, J.L.; Palacios, J.C.; Sanchez, J.B. "Hetero Diels-Alder reactions of homochiral 1,2-diaza-1,3-butadienes with diethyl azodicarboxylate under microwave irradiation. Theoretical rationale of the stereochemical outcome." *J. Org. Chem.* **1999**, *64*, pp. 6297-305.

225. Diaz-Ortiz, A.; Carrillo, J.R.; Gomez-Escalonilla, M.J.; de la Hoz, A.; Moreno, A.; Prieto, P. "First Diels-Alder reaction of pyrazolyl imines under microwave irradiation." *Synlett.* **1998**, *10*, pp. 1069-70.

226. Cado, F.; Jacquault, P.; Dozias, M.J.; Bazureau, J.P.; Hamelin, J. "Tandem conjugate carbon addition-intermolecular hetero Diels-Alder reactions using ethyl-1H-perimidine-2-acetate as a ketene aminal with heating or microwave activation." *J. Chem. Res. (S)* **1997**, pp. 176-77.

227. Garrigues, B.; Laporte, C.; Laurent, R.; Laporterie, A.; Dubac, J. "Microwave-assisted Diels-Alder reaction supported on graphite." *Liebigs Ann.* **1996**, pp. 739-41.

228. Motorina, I.A.; Fowler, F.W.; Grierson, D.S. "Intramolecular Diels-Alder reaction of N-alkyl-8-cyano-1-azadienes: a study of the Eschenmoser cycloreversion of dihydrooxazines as a route to N-alkyl-2-cyano-1-azadienes." *J. Org. Chem.* **1997**, *62*, pp. 2098-105.

229. Krishnaiah, A.; Narsaiah, B. "Studies on inverse demand hetero Diels-Alder reaction of perfluoroalkyl-2(1H)-pyridones with different dienophiles under microwave irradiation." *J. Fluorine Chem.* **2002**, *113*, pp. 133-37.

230. Shanmugasundaram, M.; Manikandan, S.; Raghunathan, R. "High chemoselectivity in microwave accelerated intramolecular domino Knoevenagel hetero Diels-Alder reactions - an efficient synthesis of pyrano[3,2c]coumarin frameworks." *Tetrahedron* **2002**, *58*, pp. 997-1003.

231. Diaz-Ortiz, A.; de la Hoz, A.; Prieto, P.; Carrillo, J.R.; Moreno, A.; Neunhoeffer, H. "Diels-Alder cycloaddition of 4,6-dimethyl-1,2,3-triazine with enamines, or their precursors, under microwave irradiation." *Synlett.* **2001**, *2*, pp. 236-37.

232. Fraile, J.M.; Garcia, J.I.; Gomez, M.A.; de la Hoz, A.; Mayoral, J.A.; Moreno, A.; Prieto, P.; Salvatella, L.; Vazquez, E. "Tandem Diels-Alder aromatization reactions of furans under unconventional reaction conditions - experimental and theoretical studies." *Eur. J. Org. Chem.* **2001**, pp. 2891-99.

233. Hijji, Y.M.; Wanene, J.; Obot, E.; Fuller, J. "Synthesis of substituted furans and substituted benzenes via microwave enhanced Diels-Alder reactions." *Fifth International Electronic Conference on Synthetic Organic Chemistry (ECSOC-5)* **2001**, E0023 (www.mdpi.net).

234. Kumareswaran, R.; Reddy, B.G.; Vankar, Y.D. "Nafion-H-catalyzed Mukaiyama aldol condensations and hetero Diels-Alder reactions using aldehydes and imines. Part 15: General synthetic methods." *Tetrahedron Lett.* **2001**, *42*, pp. 7493-95.

235. Fernandez-Paniagua, U.M.; Illescas, B.; Martin, N.; Seoane, C.; de la Cruz, P.; de la Hoz, A.; Langa, F. "Thermal and microwave-assisted synthesis of Diels-Alder adducts of (60)-fullerene with 2,3-pyrazinoquinodimethanes: characterization and electrochemical properties." *J. Org. Chem.* **1997**, *62*, pp. 3705-10.

236. Giersson, J.K.F.; Johannesdottir, J.F.; Reynisson, J. "Reactions of 1-aza-1,3-butadienes. An expedient synthesis of unsymmetrically substituted N-benzyl-1,4-dihydropyridines and N-benzyl-1,4-dihydronicotinamides." *Acta Chem. Scand.* **1997**, *51*, pp. 348-50.

237. Giguere, R.J.; Namen, A.M.; Lopez, B.O.; Arepally, A.; Ramos, D.E.; Majetich, G.; Defauw, J. "Studies on tandem ene/intramolecular Diels-Alder reactions." *Tetrahedron Lett.* **1987**, *28*, pp. 6553-56.

238. Falsone, F.S.; Kappe, C.O. "The Biginelli dihydropyrimidone synthesis using polyphosphate ester as a mild and efficient cyclocondensation/dehydration reagent." *Ark. Org. Kemi.* **2001**, *2*, pp. 1111-23.

239. Stadler, A.; Kappe, C.O. "Automated library generation using sequential microwave-assisted chemistry. Application toward the Biginelli multicomponent condensation." *J. Comb. Chem.* **2001**, *3*, pp. 624-30.

240. Stadler, A.; Kappe, C.O. "Microwave-mediated Biginelli reactions revisited. On the nature of rate and yield enhancements." *J. Chem. Soc., Perkin Trans. 1* **2000**, pp. 1363-68.

241. Yadav, J.S.; Reddy, B.V.S.; Reddy, E.J.; Ramalingam, T. "Microwave-assisted synthesis of dihydropyrimidines: improved high yielding protocol for the Biginelli reaction." *J. Chem. Res. (S)* **2000**, pp. 354-55.

242. Kappe, C.O.; Kumar, D.; Varma, R.S. "Microwave-assisted high-speed parallel synthesis of 4-aryl-3,4-dihydropyrimidin-2(1*H*)-ones using solventless Biginelli condensation protocol." *Synthesis* **1999**, pp. 1799-803.

243. Rahmouni, M.; Derdour, A.; Bazureau, J.P.; Hamelin, J. "A new access to 2,3-dihydroimidazo[1,2c]pyrimidines." *Synth. Commun.* **1996**, *26*, pp. 453-58.

244. Rahmouni, M.; Derdour, A.; Bazureau, J.P.; Hamelin, J. "A new route to pyrimido[1,6a]benzimidazoles: reactivity of activated 2-benzimidazoles with N-acyl imidates as β-dielectrophiles under microwave irradiation." *Tetrahedron Lett.* **1994**, *35*, pp. 4563-64.

245. Kappe, C.O.; Shishkin, O.V.; Uray, G.; Verdino, P. "X-ray structure, conformational analysis, enantioseparation, and determination of absolute configuration of the mitotic kinesin Eg5 inhibitor monastrol." *Tetrahedron* **2000**, *56*, pp. 1859-62.

246. Dandia, A.; Saha, M.; Taneja, H. "Synthesis of fluorinated ethyl-4-aryl-6-methyl-1,2,3,4-tetrahydropyrimidin-2-one/thione-5-carboxylates under microwave irradiation." *J. Fluorine Chem.* **1998**, *90*, pp. 17-21.

247. Saloutin, V.I.; Bughart, Y.V.; Kuzueva, O.G.; Kappe, C.O.; Chupakhin, O.N. "Biginelli condensations of fluorinated 3-oxo esters and 1,3-diketones." *J. Fluorine Chem.* **2000**, *103*, pp. 17-23.

248. Gupta, R.; Gupta, A.K.; Paul, S.; Kachroo, P.L. "Improved syntheses of some ethyl-4-aryl-6-methyl-1,2,3,4-tetrahydropyrimidin-2-one/thione-5-carboxylates by microwave irradiation." *Indian J. Chem.* **1995**, *34B*, pp. 151-52.

249. Dave, C.G.; Shah, R.D. "Gould-Jacobs type of reaction in the synthesis of thieno[3,2e]pyrimido[1,2c]pyrimidines: a comparison of classical heating vs. solvent-free microwave irradiation." *Heterocycles* **1999**, *51*, pp. 1819-26.

250. Cablewski, T.; Gurr, P.A.; Pajalic, P.J.; Strauss, C.R. "A solvent-free Jacobs-Gould reaction." *Green Chem.* **2000**, *1*, pp. 25-28.

251. Meziane, M.A.A.; Rahmouni, M.; Bazureau, J.P.; Hamelin, J. "A new route to 1-oxo-1,2-dihydropyrimido[1,6a]benzimidazole-4-carboxylates from ethyl-2-(benzimidazol-2-yl)-3-(dimethylamino)acrylate using solvent-free conditions." *Synthesis* **1998**, *7*, pp. 967-69.

252. Ferreira, V.F.; de Souza, M.C.B.V.; Cunha, A.C.; Pereira, L.O.R.; Ferreira, M.L.G. "Recent advances in the synthesis of pyrroles." *Org. Prep. Proced. Intl.* **2001**, *33*, pp. 411-54.

253. Ranu, B.C.; Hajra, A. "Synthesis of alkyl-substituted pyrroles by three-component coupling of carbonyl compound, amine and nitro-alkane/alkene on a solid surface of silica gel/alumina under microwave irradiation." *Tetrahedron* **2001**, *57*, pp. 4767-73.

254. Rao, H.S.P.; Jothilingam, S. "One-pot synthesis of pyrrole derivatives from (E)-1,4-diaryl-2-butene-1,4-diones." *Tetrahedron Lett.* **2001**, *42*, pp. 6595-97.

255. Danks, T.N. "Microwave assisted synthesis of pyrroles." *Tetrahedron Lett.* **1999**, *40*, pp. 3957-60.

256. Jolivet-Fouchet, S.; Hamelin, J.; Texier-Boullet, F.; Toupet, L.; Jacquault, P. "Novel pathway to 1-aminopyrroles and other nitrogen heterocycles from glyoxal monohydrazones and acylated active methylene compounds in solvent-free reactions under microwave irradiation." *Tetrahedron* **1998**, *54*, pp. 4561-78.

257. Ranu, B.C.; Hajra, A.; Jana, U. "Microwave-assisted synthesis of substituted pyrroles by a three-component coupling of α,β-unsaturated carbonyl compounds, amines, and nitro alkanes on the surface of silica gel." *Synlett.* **2000**, *1*, pp. 75-76.

258. Bakavoli, M.; Germaninejhad, H.; Rahimizadeh, M.; Ghassemzadeh, M.; Heravi, M.M. "Microwave assisted heterocyclization: a rapid and efficient synthesis of imidazo[4,5-b]pyridines." *Indian J. Heterocyclic Chem.* **2001**, *10*, pp. 317-18.

259. Khadilkar, B.M.; Madyar, V.R. "Scaling up of dihydropyridine ester synthesis by using aqueous hydrotrope solutions in a continuous microwave reactor." *Org. Process Res. Dev.* **2001**, *5*, pp. 452-55.

260. Ohberg, L.; Westman, J. "An efficient and fast procedure for the Hantzsch dihydropyridine synthesis under microwave conditions." *Synlett.* **2001**, pp. 1296-98.

261. Zhou, J.F.; Tu, S.J.; Feng, J.C. "One-step synthesis of pyridine derivatives from malononitrile with bisarylidenecycloalkanone under microwave irradiation." *J. Chem. Res. (S)* **2001**, pp. 268-69.

262. Sharma, U.; Ahmed, S.; Boruah, R.C. "A facile synthesis of annelated pyridines from β-formyl enamides under microwave irradiation." *Tetrahedron Lett.* **2000**, *41*, pp. 3493-95.

263. Díaz-Ortiz, A.; Carrillo, J.R.; Cossío, F.P.; Gómez-Escalonilla, M.J.; de la Hoz, A.; Moreno, A.; Prieto, P. “Synthesis of pyrazolo[3,4-b]pyridines by cycloaddition reactions under microwave irradiation.” *Tetrahedron* **2000**, *56*, p. 1569.

264. Varma, R.S.; Kumar, D. “Microwave-accelerated three-component condensation reaction on clay: solvent-free synthesis of imidazo[1,2-a] annulated pyridines, pyrazines, and pyrimidines.” *Tetrahedron Lett.* **1999**, *40*, pp. 7665-69.

265. Vega, J.A.; Vaquero, J.J.; Alvarez-Builla, J.; Ezquerra, J.; Hamdouchi, C. “A new approach to the synthesis of 2-aminoimidazo[1,2-a]pyridine derivatives through microwave-assisted N-alkylation of 2-halopyridines.” *Tetrahedron* **1999**, *55*, pp. 2317-26.

266. Cotterill, I.C.; Usyatinsky, A.Y.; Arnold, J.M.; Clark, D.S.; Dordick, J.S.; Michels, P.C.; Khmelnitsky, Y.L. “Microwave assisted combinatorial chemistry: synthesis of substituted pyridines.” *Tetrahedron Lett.* **1998**, *39*, pp. 1117-20.

267. Khadilkar, B.M.; Gaikar, V.G.; Chitnavis, A.A. “Aqueous hydrotrope solution as a safer medium for microwave enhanced Hantzsch dihydropyridine ester synthesis.” *Tetrahedron Lett.* **1996**, *37*, p. 1719.

268. Villemin, D.; Vlieghe, X. “Thiation under microwave irradiation. II: Synthesis of sulfur heterocycles.” *Sulfur Lett.* **1998**, *21*, pp. 199-203.

269. Alajarin, R.; Jordan, P.; Vaquero, J.J.; Alvarez-Builla, J. “Synthesis of unsymmetrically substituted 1,4-dihydropyridines and analogous calcium antagonists by microwave heating.” *Synthesis* **1995**, *4*, pp. 389-91.

270. Zhang, Y.W.; Shen, Z.X.; Pan, B.; Lu, X.H.; Chen, M.H. “Research on the synthesis of 1,4-dihydropyridines under microwave.” *Synth. Commun.* **1995**, *25*, pp. 857-62.

271. Khadilkar, B.M.; Gaikar, V.G.; Chitnavis, A.A. "Aqueous hydrotrope solution as a safer medium for microwave enhanced Hantzsch dihydropyridine ester synthesis." *Tetrahedron Lett.* **1995**, *36*, pp. 8083-86.

272. Khadilkar, B.M.; Chitnavis, A.A. "Rate enhancement in the synthesis of some 4-aryl-1,4-dihydropyridines using methyl-3-aminocrotonate, under microwave irradiation." *Indian J. Chem.* **1995**, *34B*, pp. 652-53.

273. Penieres, G.; Garcia, O.; Franco, K.; Hernandez, O.; Alvarez, C. "A modification to the Hantzsch method to obtain pyridines in a one pot reaction: use of bentonitic clay in a dry medium." *Heterocycl. Commun.* **1996**, *2*, pp. 359-60.

274. Paul, S.; Gupta, R.; Loupy, A. "Improved synthesis of 2-amino-3-cyanopyridines in solvent free conditions under microwave irradiation." *J. Chem. Res. (S)* **1998**, pp. 330-31.

275. Linder, M.R.; Podlech, J. "Synthesis of β-lactams from diazoketones and imines: the use of microwave irradiation." *Org. Lett.* **2001**, *3*, pp. 1849-51.

276. Kidwai, M.; Venkataramanan, R.; Kohli, S. "Alumina-supported synthesis of β-lactams using microwave." *Synth. Commun.* **2000**, *30*, pp. 989-1002.

277. Kidwai, M.; Sapra, P.; Bhushan, K.R.; Saxena, R.K.; Gupta, R.; Singh, M. "Microwave-assisted stereoselective synthesis and antibacterial activity of new fluoroquinolinyl-β-lactam derivatives." *Monatsh. Chem.* **2000**, *131*, pp. 85-90.

278. Manhas, M.S.; Banik, B.K.; Mathur, A.; Vincent, J.E.; Bose, A.K. "Vinyl-β-lactams as efficient synthons. Eco-friendly approaches via microwave assisted reactions." *Tetrahedron* **2000**, *56*, pp. 5587-601.

279. Bose, A.K.; Banik, B.K.; Mathur, C.; Wagle, D.R.; Manhas, M.S. "Polyhydroxy amino acid derivatives via β-lactams using enantiospecific approaches and microwave techniques." *Tetrahedron* **2000**, *56*, pp. 5603-19.

280. Martelli, G.; Spunta, G.; Panunzio, M. "Microwave-assisted solvent-free organic reactions: synthesis of β-lactams from 1,3-azadienes." *Tetrahedron Lett.* **1998**, *39*, pp. 6257-60.

281. Kidwai, M.; Kumar, K.; Kumar, P. "Microwave induced stereoselective synthesis and antibacterial activity of β-lactams." *J. Indian Chem. Soc.* **1998**, *75*, pp. 102-03.

282. Banik, B.K.; Jayaraman, M.; Srirajan, V.; Manhas, M.S.; Bose, A.K. "Rapid synthesis of β-lactams as intermediates for natural products via eco-friendly reactions." *J. Indian Chem. Soc.* **1997**, *74*, pp. 943-47.

283. Banik, B.K.; Manhas, M.S.; Robb, E.W.; Bose, A.K. "Environmentally benign chemistry: microwave-induced stereocontrolled synthesis of β-lactam synthons." *Heterocycles* **1997**, *44*, pp. 405-16.

284. Khajavi, M.S.; Sefidkon, F.; Hosseini, S.S.S. "Reaction of imines with trichloroacetic esters or anhydride promoted by iron carbonyl or microwave irradiation. Preparation of 3,3-dichloro-β-lactams." *J. Chem. Res. (S)* **1998**, pp. 724-25.

285. Bose, A.K.; Banik, B.K.; Manhas, M.S. "Studies on lactams. Part 98. Stereocontrol of β-lactam formation using microwave irradiation." *Tetrahedron Lett.* **1995**, *36*, pp. 213-16.

286. Kidwai, M.; Kumar, R.; Kohli, S. "Microwave-induced synthesis of nitrogen-mustard derivatives." *Indian J. Chem.* **1999**, *38B*, pp. 1132-35.

287. Azizian, J.; Soozangarzadeh, S.; Jadidi, K. "Microwave-induced one-pot synthesis of some new spiro[3H-indole-3,5'(4'H)-[1,2,4]-triazoline]-2-ones." *Synth. Commun.* **2001**, *31*, pp. 1069-73.

288. Dandia, A.; Sachdeva, H.; Singh, R. "Improved synthesis of 3-spiroindolines in dry media under microwave irradiation." *Synth. Commun.*, **2001**, *31*, pp. 1879-92.

289. Dandia, A.; Singh, R.; Sachdeva, H.; Arya, K. "Microwave assisted one pot synthesis of a series of trifluoromethyl-substituted spiro[indole-triazoles]." *J. Fluorine Chem.* **2001**, *111*, pp. 61-67.

290. Kidwai, M.; Misra, P. "Microwave-induced "solvent-free" novel technique for the synthesis of spiro[indole-pyrazole/isoxazole/pyrimidine] derivatives." *Oxidation Commun.* **2001**, *24*, pp. 287-90.

291. Dandia, A.; Sachdeva, H.; Devi, R. "Montmorillonite catalysed synthesis of novel spiro[3H-indole-3,3'-[3H-1,2,41]triazol]-2(1*H*)-ones in dry media under microwave irradiation." *J. Chem. Res. (S)* **2000**, pp. 272-75.

292. Gribble, G.W. "Recent developments in indole ring synthesis-methodology and applications." *J. Chem. Soc., Perkin Trans. 1* **2000**, pp. 1045-75.

293. Dandia, A.; Saha, M.; Taneja, H. "Improved one-pot synthesis of 3-spiroindolines under microwave irradiation." *Phosphorus Sulfur Silicon Relat. Elem.* **1998**, *139*, pp. 77-85.

294. Sridar, V. "Microwave radiation as a catalyst for chemical reactions." *Curr. Sci.* **1998**, *74*, pp. 446-50.

295. Sridar, V. "Rate acceleration of Fischer-indole cyclization by microwave irradiation." *Indian J. Chem., Sec. B* **1996**, *35*, pp. 737-38.

296. Abramovitch, R.A. "Applications of microwave energy in organic chemistry. A review." *Org. Prep. Proced. Intl.* **1991**, *23*, pp. 683-711.

297. Lipifiska, T.; Guibe-Jampel, E.; Petit, A.; Loupy, A. "2-(2-Pyridyl)indole derivatives preparation via Fischer reaction on montmorillonite K10/zinc chloride under microwave irradiation." *Synth. Commun.* **1999**, *29*, pp. 1349-54.

298. Jnaneshwara, G.K.; Bedekar, A.V.; Deshpande, V.H. "Microwave assisted preparation of isatins and synthesis of (+)-convolutamydine-A." *Synth. Commun.* **1999**, *29*, pp. 3627-33.

299. Molina, A.; Vaquero, J.J.; Garcia-Navio, J.L.; Alvarez-Builla, J. "One-pot Graebe-Ullmann synthesis of γ-carbolines under microwave irradiation." *Tetrahedron Lett.* **1993**, *34*, pp. 2673-76.

300. Ranu, B.C.; Hajra, A.; Jana, U. "Microwave-assisted simple synthesis of quinolines from anilines and alkyl vinyl ketones on the surface of silica gel in the presence of indium(III) chloride." *Tetrahedron Lett.* **2000**, *41*, pp. 531-33.

301. Sabitha, G.; Babu, R.S.; Reddy, B.V.S.; Yadav, J.S. "Microwave assisted Friedlaender condensation catalyzed by clay." *Synth. Commun.* **1999**, *29*, pp. 4303-08.

302. Ahluwalia, V.K.; Goyal, B.; Das, U. "One-pot syntheses of 5-oxo-1,4,5,6,7,8-hexahydroquinolines and pyrimido[4,5b]quinolines using microwave irradiation and ultrasound." *J. Chem. Res. (S)* **1997**, p. 266.

303. Huang, Z.Z.; Wu, L.L.; Huang, X. "Facile synthesis of 2-alkyl and 2-aryl-4-quinolones using microwave irradiation." *Chin. J. Org. Chem* **2000**, *20*, pp. 88-90.

304. Besson, T.; Dozias, M.J.; Guillard, J.; Jacquault, P.; Legoy, M.D.; Rees, C.W. "Expeditious routes to 4-alkoxyquinazoline-2-carbonitriles and thiocarbamates via N-arylimino-1,2,3-dithiazoles using microwave irradiation." *Tetrahedron* **1998**, *54*, pp. 6475-84.

305. Besson, T.; Guillard, J.; Rees, C.W. "Multistep synthesis of thiazoloquinazolines under microwave irradiation in solution." *Tetrahedron Lett.* **2000**, *41*, pp. 1027-30.

306. Seijas, J.A.; Vazquez-Tato, M.P.; Martinez, M.M. "Microwave enhanced synthesis of 4-aminoquinazolines." *Tetrahedron Lett.* **2000**, *41*, pp. 2215-17.

307. Rad-Moghadam, K.; Khajavi, M.S. "One-pot synthesis of substituted quinazoline-4(3H)-ones under microwave irradiation." *J. Chem. Res. (S)* **1998**, pp. 702-03.

308. Khajavi, M.S.; Rad-Moghadam, K.; Hazarkhani, H. "A facile synthesis of 6-substituted benzimidazo[1,2c]quinazolines under microwave irradiation." *Synth. Commun.* **1999**, *29*, pp. 2617-24.

309. Besson, T.; Rees, C.W. "New route to 4-alkoxyquinazoline-2-carbonitriles." *J. Chem. Soc., Perkin Trans. 1* **1996**, pp. 2857-60.

310. Bougrin, K.; Loupy, A.; Petit, A.; Daou, B.; Soufiaoui, M. "Novel synthesis of 2-trifluoromethylarylimidazoles on montmorillonite K-10 in a 'dry medium' under microwave irradiation." *Tetrahedron* **2001**, *57*, pp. 163-68.

311. Fresneda, P.M.; Molina, P.; Sanz, M.A. "Microwave-assisted regioselective synthesis of 2,4-disubstituted imidazoles: Nortopsentin D synthesized by minimal effort." *Synlett.* **2001**, pp. 218-21.

312. Rostamizadeh, S.; Derafshian, E. "A simple route to the preparation of benzimidazoles and benzoxazoles." *J. Chem. Res. (S)* **2001**, pp. 227-28.

313. Balalaie, S.; Arabanian, A. "One-pot synthesis of tetrasubstituted imidazoles catalyzed by zeolite HY and silica gel under microwave irradiation." *Green Chem.* **2000**, *2*, pp. 274-76.

314. Balalaie, S.; Arabanian, A.; Hashtroudi, M.S. "Zeolite HY and silica gel as new and efficient heterogenous catalysts for the synthesis of triarylimidazoles under microwave irradiation." *Monatsh. Chem.* **2000**, *131*, pp. 945-48.

315. Hashtroudi, M.S.; Nia, S.S.; Asadollahi, H.; Balalaie, S. "Microwave promoted synthesis of benzimidazole derivatives in solvent free condition." *Indian J. Heterocycl. Chem.* **2000**, *9*, pp. 307-308.

316. Usyatinsky, A.Y.; Khmelnitsky, Y.L. "Microwave-assisted synthesis of substituted imidazoles on a solid support under solvent-free conditions." *Tetrahedron Lett.* **2000**, *41*, pp. 5031-34.

317. Glas, H.; Thiel, W.R. "Microwave assisted synthesis of chiral imidazolyl and pyrazolyl alcohols." *Tetrahedron Lett.* **1998**, *39*, pp. 5509-10.

318. Bougrin, K.; Soufiaoui, M. "Novel syntheses of arylimidazoles by microwave irradiation in dry medium." *Tetrahedron Lett.* **1995**, *36*, pp. 3683-86.

319. Khajavi, M.S.; Hajihadi, M.; Naderi, R. "Synthesis of heterocyclic compounds from *o*-substituted anilines under microwave irradiation." *J. Chem. Res. (S)* **1996**, pp. 92-93.

320. Khajavi, M.S.; Hajihadi, M.; Nikpour, F. "Various syntheses of benzimidazolin-2-ones and benzimidazoline-2-thiones under microwave irradiation." *J. Chem. Res. (S)* **1996**, pp. 94-95.

321. Aghapoor, K.; Heravi, M.M.; Nooshabadi, M.A. "Synthesis of benzimidazoles in a solvent-free reaction under microwave activation." *Indian J. Chem.* **1998**, *37B*, p. 84.

322. Ben-Alloum, A.; Bakkas, S.; Soufiaoui, M. "Benzimadazoles: oxidative heterocyclization by nitrobenzene or DMSO on silica and under microwave and UV irradiation." *Tetrahedron Lett.* **1998**, *39*, pp. 4481-84.

323. Bougrin, K.; Loupy, A.; Soufiaoui, M. "Three new routes for synthesis of 1,3-azole derivatives using microwaves." *Tetrahedron* **1998**, *54*, pp. 8055-64.

324. Kamal, A.; Reddy, B.S.N.; Reddy, G.S.K. "Microwave assisted synthesis of pyrrolo[2,1c]1,4"-benzodiazepine-5,11-diones." *Synlett.* **1999**, *8*, pp. 1251-52.

325. Jolivet-Fouchet, S.; Fabis, F.; Bovy, P.; Ochsenbein, P.; Rault, S. "Novel rearrangement of pyrrolo[2,1c]1,4"-benzodiazepines into pyrrolo[2,1b]quinazolinones, analogs of alkaloid vasicinone." *Heterocycles* **1999**, *51*, pp. 1257-73.

326. Mekheimer, R.; Shaker, R.M.; Sadek, K.U.; Otto, H.H. "A novel synthesis of benzo[g"]imidazo[1,2a]pyridines: the reactivity of arylidene-1H-benzimidazole-2-acetonitrile with electron poor olefins and dimethylacetylene dicarboxylate under microwave irradiation." *Heterocycl. Commun.* **1997**, *3*, pp. 217-21.

327. Reddy, A.C.S.; Rao, P.S.; Venkataratnam, R.V. "Fluoro organics: Facile syntheses of novel 2- or 4-trifluoromethyl-1H-arylo-1,5-diazepines, oxazepines, thiazepines, 2-(1,1,1-trifluoroacetonyl) imidazoles, oxazoles, and thiazoles." *Tetrahedron* **1997**, *53*, pp. 5847-54.

328. Oussaid, B.; Moeini, L.; Martin, B.; Villemin, D.; Garrigues, B. "Improved synthesis of oxadiazoles under microwave irradiation." *Synth. Commun.* **1995**, *25*, pp. 1415-19.

329. Brain, C.T.; Paul, J.M.; Loong, Y.; Oakley, P.J. "Novel procedure for the synthesis of 1,3,4-oxadiazoles from 1,2-diacylhydrazines using polymer-supported Burgess reagent under microwave conditions." *Tetrahedron Lett.* **1999**, *40*, pp. 3275-78.

330. Brain, C.T.; Paul, J.M. "Rapid synthesis of oxazoles under microwave conditions." *Synlett.* **1999**, pp. 1642-44.

331. Lee, J.C.; Song, I.G. "Mercury(II) *p*-toluenesulfonate mediated synthesis of oxazoles under microwave irradiation." *Tetrahedron Lett.* **2000**, *41*, pp. 5891-94.

332. Oussaid, B.; Berlan, J.; Soufiaoui, M.; Garrigues, B. "Improved synthesis of oxazoline under microwave irradiation." *Synth. Commun.* **1995**, *25*, pp. 659-65.

333. Clarke, D.S.; Wood, R. "A facile one stage synthesis of oxazolines under microwave irradiation." *Synth. Commun.* **1996**, *26*, pp. 1335-40.

334. Gupta, R.; Paul, S.; Kamotra, P.; Gupta, A.K. "Rapid synthesis of S-triazolo[3,4b]-1,3,4-thiadiazoles and quinolines under microwave irradiation." *Indian J. Heterocycl. Chem.* **1997**, *7*, pp. 155-56.

335. Kidwai, M.; Kumar, P. "Microwave-induced syntheses of 6-(substituted aryl)-3-[(5-methyl-1,3,4-thiadiazol-2-yl-sulfanyl)methyl]-1,2,4-triazolo[3,4b]-1,3,4-thiadiazoles." *J. Chem. Res. (S)* **1996**, pp. 254-55.

336. Kidwai, M.; Bhushan, K.R. "A rapid one-pot synthesis of 5-substituted-2-mercapto-1,3,4-thiadiazoles using microwaves." *Indian J. Chem.* **1998**, *37B*, pp. 427-28.

337. Gupta, R.; Paul, S.; Gupta, A.K.; Kachroo, P.L. "Improved syntheses of some substituted 5,6-dihydro-S-triazolo[3,4b]-1,3,4-thiadiazoles in a microwave oven." *Indian J. Chem.* **1994**, *33B*, pp. 888-91.

338. Azizian, J.; Morady, A.V.; Jadidi, K.; Mehrdad, M.; Sarrafi, Y. "Microwave-induced one-pot synthesis of some new spiro(indoline-3,2'-thiazolidine)-2,4'-(1H)-diones and bis[(spiro)indoline-3,2'-thiazolidine]-2,4'-(1H)-diones." *Synth. Commun.* **2000**, *30*, pp. 537-42.

339. Ahluwalia, V.K.; Sharma, P.; Aggarwal, R. "Synthesis of 3-(1,3-diaryl-4,6-dioxo-2-thioxoperhydropyrimidin-5-yl)-2H-[1,4]benzothiazines." *J. Chem. Res. (S)* **1997**, pp. 16-17.

340. Ben-Alloum, A.; Bakkas, S.; Soufiaoui, M. "New synthesis pathway of 2-arylbenzothiazoles: transfer of electrons activated by microwaves." *Tetrahedron Lett.* **1997**, *38*, pp. 6395-96.

341. Besson, T.; Dozias, M.J.; Guillard, J.; Rees, C.W. "New route to 2-cyanobenzothiazoles via N-arylimino-1,2,3-dithiazoles." *J. Chem. Soc., Perkin Trans. 1* **1998**, pp. 3925-26.

342. Guillard, J.; Besson, T. "Synthesis of novel dioxinobenzothiazole derivatives." *Tetrahedron* **1999**, *55*, pp. 5139-44.

343. Beneteau, V.; Besson, T.; Rees, C.W. "Rapid synthesis of 2-cyanobenzothiazoles from N-aryliminodithiazoles under microwave irradiation." *Synth. Commun.* **1997**, *27*, pp. 2275-80.

344. Besson, T.; Guillard, J.; Rees, C.W. "Rapid synthesis of 2-cyanobenzothiazole, isothiocyanate and cyanoformanilide derivatives of dapsone." *J. Chem. Soc., Perkin Trans. 1* **2000**, pp. 563-566.

345. Bentiss, F.; Lagrenee, M.; Barbry, D. "Accelerated synthesis of 3,5-disubstituted 4-amino-1,2,4-triazoles under microwave irradiation." *Tetrahedron Lett.* **2000**, *41*, pp. 1539-41.

346. Kidwai, M.; Kohli, S.; Goel, A.K.; Dubey, M.P. "Microwave assisted synthesis and pharmacological screening of 3-(substituted phenyl)-5-methylquinolino[3,2e]-1,2,4-triazines." *Indian J. Chem.* **1998**, *38B*, pp. 1063-65.

347. Heravi, M.M.; Rahimizadeh, M.; Iravani, E.; Ghassemzadeh, M.; Aghapoor, K. "Synthesis of 3-thioxopyrido[2,3c]-1,2,4-triazine and its tricyclic derivative." *Indian J. Heterocycl. Chem.* **1999**, *9*, pp. 75-76.

348. Alterman, M.; Hallberg, A. "Fast microwave-assisted preparation of aryl and vinyl nitriles and the corresponding tetrazoles from organo-halides." *J. Org. Chem.* **2000**, *65*, pp. 7984-89.

349. Bratulescu, G. "Novel technique for one-step synthesis of 2,5-bis(alkoxycarbonyl)furans." *J. Soc. Alger. Chim.* **2000,** *10*, pp. 135-37.

350. Liao, L.; Villemin, D. "A rapid and efficient one-pot synthesis of substituted 2-(5H)-furanones under focused microwave irradiations." *J Chem. Res. (S)* **2000**, pp. 179-81.

351. Majdoub, M.; Loupy, A.; Petit, A.; Roudesli, S. "Coupling focused microwaves and solvent-free phase transfer catalysis: application to the synthesis of new furanic diethers." *Tetrahedron*, **1996**, *52*, pp. 617-28.

352. Ali, M.; Bond, S.P.; Mbogo, S.A.; McWhinnie, W.R.; Watts, P.M. "Use of a domestic microwave oven in organometallic chemistry." *J. Organomet. Chem.* **1989**, *371*, pp. 11-13.

353. Puciova, M.; Ertl, P.; Toma, S. "Synthesis of ferrocenyl-substituted heterocycles: the beneficial effect of microwave irradiation." *Collect. Czech. Chem. Commun.* **1994**, *59*, pp. 175-85.

354. CEM Corporation is very grateful to Dr. Anil Vasudevan of Abbott Laboratories (Abbott Park, IL) for performing these reactions on the *Discover*™ *System*, **2001**.

355. Kaddar, H.; Hamelin, J.; Benhaoua, H. "Microwave-assisted 1,3-dipolar cycloaddition reactions of nitrilimines and nitrile oxides." *J. Chem. Res. (S)* **1999**, pp. 718-19.

356. Lerestif, J.M.; Perrocheau, J.; Tonnard, F.; Bazareau, J.P., Hamelin, J. "1,3-Dipolar cycloaddition of imidate ylides on imino-alcohols: synthesis of new imidazolones using solvent free conditions." *Tetrahedron* **1995**, *51*, pp. 6757-74.

357. Lerestif, J.M.; Feuillet, S.; Bazareau, J.P., Hamelin, J. "Novel synthesis of protected methyl-4-hydroxy-1,2,3,4-tetrahydroisoquinoline-3-carboxylate via cleavage of functionalized dihydrooxazoles (oxazolines)." *J. Chem. Res. (S)* **1999**, pp. 32-33.

358. Ben-Alloum, A.; Bakkas, S.; Bougrin, K.; Soufiaoui, M. "Synthesis of novel spiro-rhodanine-pyrazolines by 1,3-dipolar addition of diphenylnitrilimine to some 5-arylidenerhodanines in dry medium and under microwave irradiation." *New J. Chem.* **1998**, *22*, pp. 809-12.

359. Bougrin, K.; Soufiaoui, M.; Loupy, A.; Jacquault, P. "The 1,3-dipolar addition of diphenylnitrilimine to some dipolarophiles in 'dry media' under microwaves." *New J. Chem.* **1995**, *19*, pp. 213-19.

360. Syassi, B.; Bougrin, K.; Lamira, M.; Soufiaoui, M. "One-pot synthesis of some pyrazoline and pyrazole derivatives in dry media and under microwave irradiation." *New J. Chem.* **1998**, *22*, pp. 1545-48.

361. Arrieta, A.; Carrillo, J.R.; Cossio, F.P.; Diaz-Ortiz, A.; Gomez-Escalonilla, M.J.; de la Hoz, A.; Langa, F.; Moreno, A. "Efficient tautomerization hydrazone-azomethine imine under microwave irradiation. Synthesis of (4,3') and (5,3') bipyrazoles." *Tetrahedron* **1998**, *54*, pp. 13167-180.

362. de la Cruz, P.; Díaz-Ortiz, A.; García, J.J.; Gómez-Escalonilla, M.J.; de la Hoz, A.; Langa, F. "Synthesis of new C_{60}-donor dyads by reaction of pyrazolylhydrazones with [60]fullerene under microwave irradiation." *Tetrahedron Lett.* **1999**, *40*, pp. 1587-90.

363. Diaz-Ortiz, A.; Diez-Barra, E.; de la Hoz, A.; Prieto, P.; Moreno, A. "Cycloadditions of ketene acetals under microwave irradiation in solvent-free conditions." *J. Chem. Soc., Perkin Trans. 1* **1994**, pp. 3595-98.

364. Baruah, B.; Prajapati, D.; Boruah, A.; Sandhu, J.S. "Microwave induced 1,3-dipolar cycloaddition reactions of nitrones." *Synth. Commun.* **1997**, *27*, pp. 2563-67.

365. Rigolet, S.; Goncalo, P.; Melot, J.M.; Vebrel, J. "The 1,3-dipolar cycloaddition of 3-methylene-N-substituted isoindolones and nitrones by classical and microwave techniques: reactivity and stereochemical studies." *J. Chem. Res. (S)* **1998**, pp. 686-87.

366. Diaz-Ortiz, A.; Diez-Barra, E.; de la Hoz, A.; Prieto, P.; Moreno, A.; Langa, F.; Prange, T.; Neuman, A. "Facial selectivity in cycloadditions of a chiral ketene acetal under microwave irradiation in solvent-free conditions. Configurational assignment of the cycloadducts by NOESY experiments and molecular mechanics calculations." *J. Org. Chem.* **1995**, *60*, pp. 4160-66.

367. Ondrus, V.; Orsag, M.; Fisera, L.; Pronayova, N. "Stereoselectivity of N-benzyl-C-ethoxycarbonyl nitrone cycloaddition to (S)-5-hydroxymethyl-2(5H)-furanone and its derivatives." *Tetrahedron* **1999**, *55*, pp. 10425-436.

368. Touaux, B.; Texier-Boullet, F.; Hamelin, J. "Synthesis of oximes, conversion to nitrile oxides and their subsequent 1,3-dipolar cycloaddition reactions under microwave irradiation and solvent-free reaction conditions." *Heteroatom Chem.* **1998**, *9*, pp. 351-54.

369. Touaux, B.; Klein, B.; Texier-Boullet, F.; Hamelin, J. "Synthesis in dry media coupled with microwave irradiation: application to (alkoxycarbonyl)formonitrile oxide generation and 1,3-dipolar cycloaddition." *J. Chem. Res. (S)* **1994**, pp. 116-17.

370. Syassi, B.; Bougrin, K.; Soufiaoui, M. "Addition dipolaire-1,3 des arylnitriloxydes avec quelques dipolarophiles olefiniques sur alumine en milieu sec et sous." *Tetrahedron Lett.* **1997**, *38*, pp. 8855-58.

371. Micúch, P.; Fi?era, L.; Cyranski, M.K.; Krygowski, T.M. "Reversal of stereoselectivity of Mg(II) catalysed 1,3-dipolar cycloaddition. Acceleration of cycloaddition by microwave irradiation." *Tetrahedron Lett.* **1999**, *40*, pp. 167-70.

372. de la Cruz, P.; Espíldora, E.; García, J.J.; de la Hoz, A.; Langa, F.; Martín, N.; Sánchez, L. "Electroactive 3'-(N-phenylpyrazolyl)isoxazoline[4',5':1,2][60] fullerene dyads." *Tetrahedron Lett.* **1999**, *40*, pp. 4889-92.

373. Louerat, F.; Bougrin, K.; Loupy, A.; Ochoa de Retana, A.M.; Pagalday, J.; Palacios, F. "Cycloaddition reactions of azidomethyl phosphonate with acetylenes and enamines. Synthesis of triazioles." *Heterocycles* **1998**, *48*, pp. 161-70.

374. Boruah, A.; Baruah, B.; Prajapati, D.; Sandhu, J.S.; Ghosh, A.C. "Microwave-induced 1,3-dipolar cycloaddition of 2-aroyl-aziridines." *Tetrahedron Lett.* **1996**, *37*, pp. 4203-04.

375. Langa, F.; de la Cruz, P.; de la Hoz, A.; Espildora, E.; Cossio, F.P.; Lecea, B. "Modification of regioselectivity in cycloadditions to C_{70} under microwave irradition." *J. Org. Chem.* **2000**, *65*, pp. 2499-507.

376. Diaz-Ortiz, A.; Diez-Barra, E.; de la Hoz, A.; Loupy, A. "1,3-Dipolar cycloadditions of pyridinium dicyanomethylide under microwave irradiation." *Heterocycles* **1994**, *38*, pp. 785-92.

377. de la Cruz, P.; de la Hoz, A.; Langa, F.; Illescas, B.; Martín, N. "Cycloadditions to 60 fullerene using microwave irradiation: a convenient and expeditious procedure." *Tetrahedron* **1997**, *53*, pp. 2599-608.

378. de la Cruz, P.; de la Hoz, A.; Langa, F.; Illescas, B.; Martín, N.; Seoane, C. "Microwave assisted cycloadditions to 60 fullerene." *Synth. Met.* **1997**, *86*, pp. 2283-84.

379. Kidwai, M.; Sapra, P.; Dave, B. "A facile method for nucleophilic aromatic substitution of cyclic amine." *Synth. Commun.* **2000**, *30*, pp. 4479-88.

380. Salmoria, G.V.; Dall'Oglio, E.L.; Zucco, C. "Aromatic nucleophilic substitutions under microwave irradiation." *Tetrahedron Lett.* **1998**, *39*, pp. 2471-74.

381. Abramovitch, R.A.; Abramovitch, D.A.; Iyanar, K.; Tamareselvy, K. "Application of microwave energy to organic synthesis: improved technology." *Tetrahedron Lett.* **1991**, *32*, pp. 5251-54.

382. Caddick, S. "Microwave assisted organic reactions." *Tetrahedron* **1995**, *51*, pp. 10403-432.

383. Heber, D.; Stoyanov, E.V. "A microwave assisted nucleophilic substitution of 4-hydroxy-6-methyl-2(1H)-pyridones." *Synlett.* **1999**, *11*, pp. 1747-48.

384. Yadav, J.S.; Reddy, B.V.S. "CsF-Al_2O_3 mediated rapid condensation of phenols with aryl halides: comparative study of conventional heating vs. microwave irradiation." *New J. Chem.* **2000**, *24*, pp. 489-91.

385. Jiang, Y.L.; Pang, J.; Yuan, Y.C. "A novel preparation of *o*-ethoxyphenol from *o*-chlorophenol in the presence of phase transfer catalysts under microwave irradiation." *Chin. Chem. Lett.* **1994**, *5*, pp. 29-30.

386. Sagar, A.D.; Patil, D.S.; Bandgar, B.P. "Microwave assisted synthesis of triaryl cyanurates." *Synth. Commun.* **2000**, *30*, pp. 1719-23.

387. Bansal, V.; Kanodia, S.; Thapliyal, P.C.; Khanna, R.N. "Microwave induced selective bromination of 1,4-quinones and coumarins." *Synth. Commun.* **1996**, *26*, pp. 887-92.

388. Chakrabarty, M.; Basak, R.; Ghosh, N. "Microwave-assisted Michael reactions of 3-(2'-nitrovinyl) indole with indoles on TLC-grade silica gel. A new, facile synthesis of 2,2-bis(3'-indolyl)nitroethanes." *Tetrahedron Lett.* **2001**, *42*, pp. 3913-15.

389. de la Hoz, A.; Diaz-Ortiz, A.; Gomez, M.V.; Mayoral, J.A.; Moreno, A.; Sanchez-Migallon, A.M.; Vazquez, E. "Preparation of α- and β-substituted alanine derivatives by α-amidoalkylation or Michael addition reactions under heterogeneous catalysis assisted by microwave irradiation." *Tetrahedron* **2001**, *57*, pp. 5421-28.

390. Ranu, B.C.; Guchhait, S.K.; Ghosh, K.; Patra, A. "Construction of bicyclo[2.2.2]octanone systems by microwave-assisted solid phase Michael addition followed by Al_2O_3-mediated intramolecular aldolisation. An eco-friendly approach." *Green Chem.* **2000**, *2*, pp. 5-6.

391. Kayama, H.; Sawaguchi, M.; Nagata, C. "Acceleration of Michael addition reaction by microwave irradiation in the presence of metal acetylacetonate catalysts." *J. Chem. Soc. Japan, Chem. Indus. Chem.* **1999**, pp. 145-48.

392. Baruah, B.; Boruah, A.; Prajapati, D.; Sandhu, J.S. "$BiCl_3$ or CdI_2 catalyzed Michael additions of 1,3-dicarbonyl compounds under microwave irradiations." *Tetrahedron Lett.* **1997**, *38*, pp. 1449-50.

393. Ibrahim-Ouali, M.; Sinibaldi, M.E.; Troin, Y.; Gardette, D.; Gramain, J.C. "Synthesis of β-enaminoesters and lactams by Michael addition of N-benzylaniline to new allenic esters and lactams." *Synth. Commun.* **1997**, *27*, pp. 1827-48.

394. Ranu, B.C.; Saha, M.; Bhar, S. "Microwave-assisted Michael addition of cycloalkenones and substituted enones on the surface of alumina in dry media." *Synth. Commun.* **1997**, *27*, pp. 621-26.

395. Soriente, A.; Spinella, A.; De Rosa, M.; Giordano, M.; Scettri, A. "Solvent-free reaction under microwave irradiation: A new procedure for Eu(III)-catalyzed Michael addition of 1,3-dicarbonyl compounds." *Tetrahedron Lett.* **1997**, *38*, pp. 289-90.

396. Boruah, A.; Baruah, M.; Prajapati, D.; Sandhu, J.S. "Cerium catalyzed Michael addition of 1,3-dicarbonyl compounds under microwave irradiation." *Synth. Commun.* **1998**, *28*, pp. 653-58.

397. Michaud, D.; Texier-Boullet, F.; Hamelin, J. "Michael monoaddition of nitromethane on gem-diactivated alkenes in dry media coupled with microwave irradiation." *Tetrahedron Lett.* **1997**, *38*, pp. 7563-64.

398. Boruah, A.; Baruah, B.; Prajapati, D.; Sandhu, J.S. "Michael reaction in the solid state under microwave irradiations." *Chem. Lett.* **1996**, pp. 965-66.

399. Sviridova, L.A.; Golubeva, G.A. "New method for the direct cyanoethylation of pyrazole derivatives." *Chem. Heterocycl. Compd.* **1999**, *35*, p. 245.

400. Moghaddam, F.M.; Mohammadi, M.; Hosseinnia, A. "Water promoted Michael addition of secondary amines to α,β-unsaturated carbonyl compounds under microwave irradiation." *Synth. Commun.* **2000**, *30*, pp. 643-50.

401. Romanova, N.N.; Gravis, A.G.; Leshcheva, I.F.; Bundel, Y.G. "1,2-Asymmetric induction in nucleophilic Michael addition reactions of amines under microwave irradiation." *Mendeleev Commun.* **1998**, pp. 147-48.

402. Romanova, N.N.; Gravis, A.G.; Shaidullina, G.M.; Leshcheva, L.F.; Bundel, Y.G. "The application of microwave irradiation to the Michael synthesis of esters of amino acids." *Mendeleev Commun.* **1997**, pp. 235-36.

403. Steinreiber, A.; Stadler, A.; Mayer, S.F.; Faber, K.; Kappe, C.O. "High-speed microwave-promoted Mitsunobu inversions. Application toward the deracemization of sulcatol." *Tetrahedron Lett.* **2001**, *42*, pp. 6283-86.

404. Jun, C.H.; Lee, H.; Hong, J.B. "Chelation-assisted intermolecular hydroacylation: direct synthesis of ketone from aldehyde and 1-alkene." *J. Org. Chem.* **1997**, *62*, pp. 1200-01.

405. Borah, H.N.; Boruah, R.C.; Sandhu, J.S. "Microwave-induced one-pot synthesis of N-(carboxyalkyl)-maleimides and –phthalimides." *J. Chem. Res. (S)* **1998**, pp. 272-73.

406. Chandrasekhar, S.; Takhi, M.; Uma, G. "Solvent free N-alkyl and N-arylimide preparation from anhydrides catalyzed by $TaCl_5$-silica gel." *Tetrahedron Lett.* **1997**, *38*, pp. 8089-92.

407. Seijas, J.A.; Vázquez-Tato, M.P.; Martínez, M.M.; Núñez-Corredoira, G. "Direct synthesis of imides from dicarboxylic acids using microwaves." *J. Chem. Res. (S)* **1999**, pp. 420-21.

408. Bose, A.K.; Jayaraman, M.; Okawa, A.; Bari, S.S. "Microwave-assisted rapid synthesis of amino-lactams." *Tetrahedron Lett.* **1996**, *37*, pp. 6989-92.

409. Vidal, T.; Petit, A.; Loupy, A.; Gedye, R.N. "Re-examination of microwave-induced synthesis of phthalimides." *Tetrahedron* **2000**, *56*, pp. 5473-78.

410. Khajavi, M.S.; Nikpour, F.; Hajihadi, M. "Microwave irradiation promoted reactions of anhydrides with isocyanates. Preparation of N-substituted phthalimides." *J. Chem. Res. (S)* **1996**, pp. 96-97.

411. Kidwai, M.; Kumar, R. "Microwave-assisted synthesis of novel 1,3,4-thiadiazolyl-substituted-1,2,4-tetrazines, pyridazinones, 1,2,4-triazoles, 4-thiazolidinones, oxazoles, and thiazoles." *Gazz. Chim. Ital.* **1997**, *127*, pp. 263-68.

412. Mojtahedi, M.M.; Saidi, M.R.; Bolourtchian, M. "A novel method for the synthesis of disubstituted ureas and thioureas under microwave irradiaton." *J. Chem. Res. (S)* **1999**, pp. 710-11.

413. Marquez, H.; Plutin, A.M.; Rodriguez, Y.; Perez, E.; Loupy, A. "Efficient synthesis of 1-(4'-methylbenzoyl)-3,3-diethylthiourea under microwave irradiation using potassium fluoride on alumina." *Synth. Commun.* **2000**, *30*, pp. 1067-74.

414. Kidwai, M.; Goel, Y.; Kumar, R. "Microwave-assisted synthesis and antifungal activity of 1,2,4-triazine, 1,2,4-triazole, tetrazole, and pyrazole derivatives." *Indian J. Chem.* **1998**, *37B*, pp. 174-79.

415. Williams, L. "Thin layer chromatography as a tool for reaction optimisation in microwave assisted synthesis." *J. Chem. Soc., Chem. Commun.* **2000**, pp. 435-36.

416. Marquez, H.; Perez, E.R.; Plutin, A.M.; Morales, M.; Loupy, A. "Synthesis of 1-benzoyl-3-alkylthioureas by transamidation under microwave in dry media." *Tetrahedron Lett.* **2000**, *41*, pp. 1753-56.

417. Gadhwal, S.; Dutta, M.P.; Boruah, A.; Prajapati, D.; Sandhu, J.S. "Zeolite-HY: a selective and efficient catalyst for the synthesis of amides under microwave irradiations." *Indian J. Chem.* **1998**, *37B*, pp. 725-27.

418. Baldwin, B.W.; Hirose, T.; Wang, Z.H. "Improved microwave oven synthesis of amides and imides promoted by imidazole; convenient transport agent preparation." *J. Chem. Soc., Chem. Commun.* **1996**, pp. 2669-70.

419. Dayal, B.; Rapole, K.R.; Patel, C.; Pramanik, B.N. "Microwave-induced rapid synthesis of sarcosine conjugated bile acids." *Bioorg. Med. Chem. Lett.* **1995**, *5*, pp. 1301-06.

420. Linares, R.M.; Ayala, J.H.; Afonso, A.M.; Gonzalez, V. "Quantitative analysis of biogenic amines by high-performance thin-layer chromatography utilizing a fibre optic fluorescence detector." *Anal. Chem.* **1998**, *31*, pp. 475-89.

421. Kidwai, M.; Kumar, R.; Srivastava, A.; Gupta, H.P. "Microwave-assisted synthesis of novel 1,3,4-thiadiazolyl-substituted-1,2,4-triazines as potential antitubercular agents." *Bioorg. Chem.* **1998**, *26*, pp. 289-94.

422. Iranpoor, N.; Zeynizadeh, B. "Microwave-promoted trifluoroacetylation of amines with $TiO(CF_3CO_2)_2$." *J. Chem. Res. (S)* **1999**, pp. 124-25.

423. Kalita, D.J.; Borah, R.; Sarma. J.C. "A selective catalytic method of enol-acetylation under microwave irradiation." *J. Chem. Res. (S)* **1999**, 404-05.

424. Deka, N.; Mariotte, A.M.; Boumendjel, A. "Microwave-mediated solvent-free acetylation of deactivated and hindered phenols." *Green Chem.* **2001**, *3*, pp. 263-64.

425. Oussaid, A.; Pentek, E.; Loupy, A. "Selective alkylations of 2-naphthol using solvent-free conditions under microwave irradiation." *New J. Chem.* **1997**, *21*, pp. 1339-45.

426. Vanelle, P.; Gellis, A.; Kaafarani, M.; Maldonado, J.; Crozet, M.P. "Fast electron transfer C-alkylation of 2-nitropropane anion under microwave irradiation." *Tetrahedron Lett.* **1999**, *40*, pp. 4343-46.

427. Bansal, V.; Singh, P.K.; Khanna, R.N. "New synthesis of 1,3-diarylpropane-1,3-diones." *Indian J. Chem.* **1996**, *35B*, pp. 586-87.

428. Deng, R.; Wang, Y.; Jiang, Y. "Solid-liquid phase transfer catalytic synthesis. X. The rapid alkylation of ethyl acetoacetate under microwave irradiation." *Synth. Commun.* **1994**, *24*, pp. 111-15.

429. Abramovitch, R.A.; Shi, Q.; Bogdal, D. "Microwave-assisted alkylations of activated methylene groups." *Synth. Commun.* **1995**, *25*, pp. 1-8.

430. Kumar, H.M.S.; Reddy, B.V.S.; Reddy, E.J.; Yadav, J.S. "Microwave-assisted eco-friendly synthesis of 2-alkylated hydroquinones in dry media." *Green Chem.* **1999**, *1*, pp. 141-42.

431. Abdallah-El Ayoubi, S.; Toupet, L.; Texier-Boullet, F.; Hamelin, J. "New route to functionalized cyclohexenes in solvent-free conditions from enamino ketones and oxo alkenes." *Synthesis* **1999**, *7*, pp. 1112-16.

432. Ruault, P.; Pilard, J.F.; Touaux, B.; Texier-Boullet, F.; Hamelin, J. "Rapid generation of amines by microwave irradiation of ureas dispersed on clay." *Synlett.* **1994**, *11*, pp. 935-36.

433. Mojtahedi, M.M.; Sharifi, A.; Mohsenzadeh, F.; Saidi, M.R. "Microwave-assisted aminomethylation of electron-rich compounds under solvent-free condition." *Synth. Commun.* **2000**, *30*, pp. 69-72.

434. Villemin, D.; Sauvaget, F. "Dry synthesis under microwave irradiation: a rapid and efficient coupling of naphthols." *Synlett.* **1994**, *6*, pp. 435-36.

435. Cado, F.; Di-Martino, J.L.; Jacquault, P.; Bazureau, J.P.; Hamelin, J. "Amidine-enediamine tautomerism: addition of isocyanates to 2-substituted 1H-perimidines. Some syntheses under microwave irradiation." *Bull. Soc. Chim. Fr.* **1996**, *133*, pp. 587-95.

436. Carrillo, J.R.; Diaz-Ortiz, A.; de la Hoz, A.; Gomez-Escalonilla, M.J.; Moreno, A.; Prieto, P. "The effect of focused microwaves on the reaction of ethyl N-trichloroethylidenecarbamate with pyrazole derivatives." *Tetrahedron* **1999**, *55*, pp. 9623-30.

437. Kundu, M.K.; Mukherjee, S.B.; Balu, N.; Padmakumar, R. "Microwave mediated extensive rate enhancement of Baylis-Hillman reaction." *Synlett.* **1994**, *6*, p. 444.

438. Garrigues, B.; Laurent, R.; Laporte, C.; Laporterie, A.; Dubac, J. "Microwave-assisted carbonyl Diels-Alder and carbonyl-ene reactions supported on graphite." *Liebigs Ann.* **1996**, pp. 743-44.

439. Almena, I.; Diaz-Ortiz, A.; Diez-Barra, E.; de la Hoz, A.; Loupy, A. "Solvent-free benzylations of 2-pyridone. Regiospecific N- or C-alkylation." *Chem. Lett.* **1996**, pp. 333-34.

440. Torchy, S.; Barbry, D. "N-alkylation of amines under microwave irradiation: modified Eschweiler-Clarke reaction." *J. Chem. Res. (S)* **2001**, pp. 292-93.

441. Fan, X.J.; You, J.M.; Jiao, T.Q.; Tan, G.Z.; Yu, X.D. "Rapid N-alkylation of carbazole, phenothiazine, and acridone under microwave irradiation." *Org. Prep. Proc. Intl.* **2000**, *32*, pp. 284-87.

442. Bogdal, D.; Pielichowski, J.; Jaskot, K. "Remarkable fast N-alkylation of azaheterocycles under microwave irradition in dry media." *Heterocycles* **1997**, *45*, pp. 715-22.

443. Bogdal, D.; Pielichowski, J.; Jaskot, K. "New synthesis method of N-alkylation of carbazole under microwave irradition in dry media." *Synth. Commun.* **1997**, *27*, pp. 1553-60.

444. Bogdal, D.; Pielichowski, J.; Boron, A. "Remarkable fast microwave-assisted N-alkylation of phthalimide in dry media." *Synlett.* **1996**, *37*, pp. 873-74.

445. Ding, J.; Gu, H.; Wen, J.; Lin, C. "Dry reaction under microwave: N-alkylation of saccharin on silica gel." *Synth. Commun.* **1994**, *24*, pp. 301-03.

446. Gupta, R.; Paul, S.; Gupta, A.K.; Kachroo, P.L.; Dandia, A. "Opening of oxirane ring with N-nucleophiles under microwave irradiation." *Indian J. Chem.* **1997**, *36B*, pp. 281-83.

447. Sabitha, G.; Reddy, B.V.S.; Abraham, S.; Yadav, J.S. "Microwave promoted synthesis of aminoalcohols in dry media." *Green Chem.* **1999**, *1*, pp. 251-52.

448. Lindström, U.M.; Olofsson, B.; Somfai, P. "Microwave-assisted aminolysis of vinylepoxides." *Tetrahedron Lett.* **1999**, *40*, pp. 9273-76.

449. Yadav, J.S.; Reddy, B.V.S. "Microwave-assisted efficient synthesis of N-arylamines in dry media." *Green Chem.* **2000**, *2*, pp. 115-16.

450. Jaisinghani, H.G.; Khadilkar, B.M. "Microwave-assisted, highly efficient solid state N- and S-alkylation." *Synth. Commun.* **1999**, *29*, pp. 3693-98.

451. Adamczyk, M.; Rege, S. "Microwave assisted sulfopropylation of N-heterocycles using 1,3-propane sultone." *Tetrahedron Lett.* **1998**, *39*, pp. 9587-88.

452. Wang, C.D.; Lu, J.; Shi, X.Z.; Feng, Y.H. "Synthesis of thiosemicarbazones under microwave irradiation." *Synth. Commun.* **1999**, *29*, pp. 3057-61.

453. Molina, P.; Fresneda, P.M.; Delgado, S. "Iminophosphorane-mediated synthesis of the alkaloid cryptotackieine." *Synthesis* **1999**, *2*, pp. 326-29.

454. Forfar, I.; Cabildo, P.; Claramunt, R.M.; Elguero, J. "Synthesis of 3-(1-adamantyl)pyrazole and 3,5-di(1-adamantyl)pyrazole in a microwave oven." *Chem. Lett.* **1994**, pp. 2079-80.

455. Almena, I.; Diez-Barra, E.; de la Hoz, A.; Riuz, J.; Sanchez-Migallon, A.; Elguero, J. "Alkylation and arylation of pyrazoles under solvent-free conditions: conventional heating versus microwave irradiation." *J. Heterocycl. Chem.* **1998**, *35*, pp. 1263-68.

456. Abenhaim, D.; Diez-Barra, E.; de la Hoz, A.; Loupy, A.; Sanchez, M.A. "Selective alkylation of 1,2,4-triazole and benzotriazole in the absence of solvent." *Heterocycles* **1994**, *38*, pp. 793-802.

457. Stankovicova, H.; Gasparova, R.; Lacova, M.; Chovancova, J. "Reaction of 4-oxochromene-3-carboxaldehydes with primary amides and benzotriazole or 1H-1,2,4-triazole." *Collect. Czech. Chem. Commun.* **1997**, *62*, pp. 793-802.

458. Kornet, M.J. "Microwave synthesis and anticonvulsant activity of new 3-benzyl-1,2,3-benzotriazin-4(3H)ones." *J. Heterocycl. Chem.* **1997**, *34*, pp. 1391-93.

459. Kornet, M.J.; Shackleford, G. "Microwave synthesis and anticonvulsant activity of new 2-benzyl-1(2H)-phthalazinones." *J. Heterocycl. Chem.* **1999**, *36*, pp. 1095-96.

460. Jiang, Y.L.; Hu, Y.Q.; Feng, S.Q.; Wu, J.S.; Wu, Z.W.; Yuan, Y.C. "Facile N-alkylation of anilines with alcohols over Raney nickel under microwave irradiation." *Synth. Commun.* **1996**, *26*, pp. 161-64.

461. Barbry, D.; Torchy, S. "Fast N-methylation of amines under microwave irradiation." *Synth. Commun.* **1996**, *26*, pp. 3919-22.

462. Stone-Elander, S.A.; Elander, N.; Thorell, J.O.; Solas, G.; Svennebrink, J. "A single-mode microwave cavity for reducing radiolabeling reaction times, demonstrated by alkylation with (11C)alkyl halides." *J. Labelled Compd. Radiopharm.* **1994**, *34*, pp. 949-60.

463. Kidwai, M.; Kumar, P.; Goel, Y.; Kumar, K. "Microwave assisted synthesis of 5-methyl-1,2,4-thiadiazol-2-yl/thiotetrazol-1-yl substituted pyrazoles, 2-azetidinones, 4-thiazolidinones, benzopyran-2-ones, and 1,3,4-oxadiazoles." *Indian J. Chem.* **1997**, *36B*, pp. 175-79.

464. Sharifi, A.; Mirzaei, M.; Nairni-Jamal, M.R. "Solvent-free aminoalkylation of phenols and indoles assisted by microwave irradiation." *Monatsh. Chem.* **2001**, *132*, pp. 875-80.

465. Kidwai, M.; Misra, P.; Kumar, R.; Saxena, R.K.; Gupta, R.; Bradoo, S. "Microwave assisted synthesis and antibacterial activity of new quinolone derivatives." *Monatsh. Chem.* **1998**, *129*, pp. 961-66.

466. Kidwai, M.; Kohli, S. "Environmentally co-friendly thiolation of 1,4-naphthoquinone." *Indian J. Chem.* **1998**, *37B*, pp. 1294-95.

467. Kidwai, M.; Bhushan, K.R.; Misra, P. "A rapid and cheap synthesis of cephalosporins." *Chem. Lett.* **1999**, pp. 487-88.

468. Kumar, P.; Gupta, K.C. "Microwave assisted synthesis of S-trityl and S-acylmercapto alkanols, nucleosides and their deprotection." *Chem. Lett.* **1996**, pp. 635-36.

469. Soukri, M.; Guillaumet, G.; Besson, T.; Aziane, D.; Aadil, M.; Essassi, E.M.; Akssira, M. "Synthesis of novel 5a,10,14b,15-tetraaza-benzo[a]indeno[1,2-c]anthracen-5-one and benzimidazo[1,2-c]quinazoline derivatives under microwave irradiation." *Tetrahedron Lett.* **2000**, *41*, pp. 5857-60.

470. Vanden Eynde, J.J.; Mailleux, I. "Quaternary ammonium salt-assisted organic reactions in water: alkylation of phenols." *Synth. Commun.* **2001**, *31*, pp. 1-7.

471. Lima, L.M.; Barreiro, E.J.; Fraga, C.A.M. "O-alkylation of bioactive phthalimide derivatives under microwave irradiation in dry media." *Synth. Commun.* **2000**, *30*, pp. 3291-306.

472. Chatti, S.; Bortolussi, M.; Loupy, A. "Synthesis of diethers derived from dianhydrohexitols by phase transfer catalysis under microwave." *Tetrahedron Lett.* **2000**, *41*, pp. 3367-70.

473. Mitra, A.K.; De, A.; Karchaudhuri, N. "Microwave enhanced synthesis of aromatic methyl ether." *Indian J. Chem. Sect. B* **2000**, *39*, pp. 387-89.

474. Motorina, I.A.; Parly, F.; Grierson, D.S. "Selective O-allylation of amido alcohols on solid support." *Synlett.* **1996**, *4*, pp. 389-91.

475. Khadilkar, B.M.; Bendale, P.M. "Microwave enhanced synthesis of epoxypropoxyphenols." *Synth. Commun.* **1997**, *27*, pp. 2051-56.

476. Wang, J.X.; Zhang, Y.; Huang, D.; Hu, Y. "Solid-liquid phase-transfer catalytic synthesis of chiral glycerol sulfide ethers under microwave irradiation." *J. Chem. Res. (S)* **1998**, pp. 216-17.

477. Pchelka, B.; Plenkiewicz, J. "Microwave-promoted synthesis of 1-aryloxy-3-alkylamino-2-propanols." *Org. Prep. Proced. Int.* **1998**, *30*, pp. 87-89.

478. Bagnell, L.; Cablewski, T.; Strauss, C.R. "A catalytic symmetrical etherification." *J. Chem. Soc., Chem. Commun.* **1999**, pp. 283-84.

479. Zadmard, R.; Aghapoor, K.; Bolourtchian, M.; Saidi, M.R. "Solid composite copper-copper chloride assisted alkylation of naphthols promoted by microwave irradiation." *Synth. Commun.* **1998**, *28*, pp. 4495-99.

480. Wang, Z.Y.; Shi, H.J.; Shi, H.X.; Zhang, Z.Y. "The synthesis of aryloxycarboxylic acid under microwave, solid base as a support." *Chin. Chem. Lett.* **1996**, *7*, pp. 527-30.

481. Elder, J.W.; Holtz, K.M. "Microwave microscale organic experiments." *J. Chem. Educ.* **1996**, *73*, pp. A104-A105.

482. Khalafi-Nezhad, A.; Hashemi, A. "Efficient synthesis of sodium aryloxymethanesulfonates using microwave irradiation." *J. Chem. Res. (S)* **1999**, pp. 720-21.

483. Pang, J.; Xi, Z.; Cao, G.; Yuan, Y. "Phase transfer catalyzed synthesis of *o*-ethoxyphenol under microwave irradiation." *Synth. Commun.* **1996**, *26*, pp. 3425-29.

484. Kidwai, M.; Kumar, P.; Kohli, S. "Microwave-induced selective alkoxylation of 1,4-naphthoquinones." *J. Chem. Res. (S)* **1997**, pp. 24-25.

485. Bansal, V.; Sharma, J.; Khanna, R.N. "Microwave-induced monohydroxymethylation and monoalkoxylation of 1,4-naphthoquinones." *J. Chem. Res. (S)* **1998**, pp. 720-21.

486. Wang, J.X.; Zhang, M.; Xing, Z.; Hu, Y. "Synthesis of aromatic ethers without organic solvent and inorganic carrier under microwave irradiation." *Synth. Commun.* **1996**, *26*, pp. 301-05.

487. Bratulescu, G. "Organic synthesis in the absence of solvent and absorbent support via microwave activation. Application to the synthesis of azoxy ethers." *Rev. Roum. Chim.* **1998**, *43*, pp. 1153-56.

488. Bogdal, D.; Pielichowski; J.; Boron, A. "New synthetic method of aromatic ethers under microwave irradiation in dry media." *Synth. Commun.* **1998**, *28*, pp. 3029-39.

489. Wang, J.X.; Zhang, M.; Hu, Y. "Synthesis of 8-quinolinyl ethers under microwave irradiation." *Synth. Commun.* **1998**, *28*, pp. 2407-13.

490. Reddy, Y.T.; Rao, M.K.; Rajitha, B. "Facile synthesis of desyl ethers under phase transfer catalytic and solvent free microwave conditions." *Indian J. Heterocycl. Chem.* **2000**, *101*, pp. 73-74.

491. Majdoub, M.; Loupy, A.; Petit, A.; Roudesli, S. "Coupling focused microwaves and solvent-free phase transfer catalysis: application to the synthesis of new furanic diethers." *Tetrahedron* **1996**, *52*, pp. 617-28.

492. Stadler, A.; Kappe, C.O. "The effect of microwave irradiation on carbodiimide-mediated esterifications on solid support." *Tetrahedron* **2001**, *57*, pp. 3915-20.

493. Fan, X.J.; You, J.M.; Jiao, T.Q.; Tan, G.Z.; Yu, X.D. "Esterification by microwave irradiation on activated carbon." *Org. Prep. Proc. Intl.* **2000**, *32*, pp. 287-90.

494. Kabza, K.G.; Chapados, B.R.; Gestwicki, J.E.; McGrath, J.L. "Microwave-induced esterification using heterogeneous acid catalyst in a low dielectric constant medium." *J. Org. Chem.* **2000**, *65*, pp. 1210-14.

495. Mitra, A.K.; De, A.; Karchaudhuri, N. "Microwave enhanced esterification of α,β-unsaturated acids." *Indian J. Chem. Sect. B* **2000**, *39*, pp. 311-12.

496. Lami, L.; Casal, B.; Cuadra, L.; Merino, J.; Alvarez, A.; Ruiz-Hitzky, E. "Synthesis of 2,4-D-ester herbicides - new routes using inorganic solid supports." *Green Chem.* **1999**, *1*, pp. 199-204.

497. Chemat, F.; Poux, M.; Galema, S.A. "Esterification of stearic acid by isomeric forms of butanol in a microwave oven under homogeneous and heterogeneous reaction conditions." *J. Chem. Soc., Perkin Trans. 2* **1997**, pp. 2371-74.

498. Jiang, Y.L.; Yuan, Y.C. "The tribromolanthanoids ($LnBr_3$) catalyzed reactions of benzyl alkyl ethers and carboxylic acids promoted by microwave irradiation." *Synth. Commun.* **1994**, *24*, pp. 1045-48.

499. Kwon, P.S.; Kim, J.K.; Kwon, T.W.; Kim, Y.H.; Chung, S.K. "Microwave-irradiated acetylation and nitration of aromatic compounds." *Bull. Korean Chem. Soc.* **1997**, *18*, pp. 1118-19.

500. Dasgupta, A.; Thompson, W.C.; Malik, S. "Use of microwave irradiation for rapid synthesis of perfluorooctanoyl derivatives of fatty alcohols, a new derivative for gas chromatograpy-mass spectrometric and fast atom bombardment mass spectrometric study." *J. Chromatogr. A* **1994**, *685*, pp. 279-85.

501. Gelo-Pujic, M.; Guibe-Jampel, E.; Loupy, A.; Galema, S.A.; Mathe, D. "Lipase-catalyzed esterification of some α-D-glucopyranosides in dry media using focused microwave irradiation." *J. Chem. Soc., Perkin Trans. 1* **1996**, pp. 2777-80.

502. Herradon, B.; Morcuende, A.; Valverde, S. "Microwave accelerated organic transformations: dibutylstannylene acetal mediated selective acylation of polyols and amino alcohols using catalytic amounts of dibutyltin oxide. Influence of the solvent and power output on the selectivity." *Synlett.* **1995**, *5*, pp. 455-58.

503. Morcuende, A.; Valverde, S.; Herradon, B. "Rapid formation of dibutylstannylene acetals from polyhydroxylated compounds under microwave heating. Application to the regioselective protection of polyols and to a catalytic tin-mediated benzoylation." *Synlett.* **1994**, *1*, pp. 89-91.

504. Morcuende, A.; Ors, M.; Valverde, S.; Herradon, B. "Microwave-promoted transformations: fast and chemoselective N-acylation of amino alcohols using catalytic amounts of dibutyltin oxide. Influence of the power output and the nature of the acylating agent on the selectivity." *J. Org. Chem.* **1996**, *61*, pp. 5264-70.

505. Bram, G.; Loupy, A.; Majdoub, M. "Microwave irradiation plus solid-liquid phase transfer catalysis without solvent: further improvement in anionic activation." *Synth. Commun.* **1990**, *20*, pp. 125-29.

506. Bram, G.; Loupy, A.; Majdoub, M.; Gutierrez, E.; Ruiz-Hitzky, E. "Alkylation of potassium acetate in 'dry media'. Thermal activation in commercial microwave ovens." *Tetrahedron* **1990**, *46*, pp. 5167-76.

507. Loupy, A.; Petit, A.; Ramdani, M.; Yvanaeff, C.; Majdoub, M.; Labiad, B.; Villemin, D. "The synthesis of esters under microwave irradiation using dry-media conditions." *Can. J. Chem.* **1993**, *71*, pp. 90-95.

508. Limousin, C.; Cléophax, J.; Loupy, A.; Petit, A. "Synthesis of benzoyl and dodecanoyl derivatives from protected carbohydrates under focused microwave irradiation." *Tetrahedron* **1998**, *54*, pp. 13567-578.

509. Balaji, B.S.; Chanda, B.M. "Simple and high yielding syntheses of β-keto esters catalysed by zeolites." *Tetrahedron* **1998**, *54*, pp. 13237-252.

510. Limousin, C.; Olesker, A.; Cléophax, J.; Petit, A.; Loupy, A.; Lukacs, G. "Halogenation of carbohydrates by triphenylphosphine complex reagents in highly concentrated solution under microwave activation or conventional heating." *Carbohydr. Res.* **1998**, *312*, pp. 23-31.

511. Kidwai, M.; Kohli, S.; Kumar, P. "Rapid side-chain chlorination of heterocyclic compounds using focused microwave irradiation." *J. Chem. Res. (S)* **1998**, pp. 586-87.

512. Kidwai, M.; Sapra, P.; Bhushan, K.R. "Fluorination of 2-chloro-3-formylquinolines using microwaves." *Indian J. Chem.* **1999**, *38B*, pp. 114-15.

513. Dolci, L.; Dolle, F.; Valette, H.; Vaufrey, F.; Chantal, F.; Bottlaender, M.; Crouzel, C. "Synthesis of a fluorine-18 labeled derivative of epibatidine for *in vivo* nicotinic acetylcholine receptor PET imaging." *Bioorg. Med. Chem.* **1999**, *7*, pp. 467-79.

514. Johnstroem, P.S.; Stone-Elander, S. "Strategies for reducing isotopic dilution in the synthesis of ^{18}F-labeled polyfluorinated ethyl groups." *Appl. Radiat. Isot.* **1996**, *47*, pp. 401-07.

515. Singh, S.; Jimbow, K.; Kumar, P.; McEwan, A.J.; Wiebe, L.I. "Synthesis and radioiodination of 3-(E)-(2-iodovinyl)-N-acetyl-4-cysteaminylphenol, a putative tyrosinase for imaging neural crest tumors." *J. Labelled Compd. Radiopharm.* **1998**, *41*, pp. 355-61.

516. Chirakal, R.; Firnau, G.; Garnett, S.; McCarry, B.; Lonergan, M. "Base-mediated decomposition of mannose triflate during the synthesis of 2-deoxy-2-^{18}F-fluoro-D-glucose." *Appl. Radiat. Isot.* **1995**, *46*, pp. 149-55.

517. Taylor, M.D.; Roberts, A.D.; Nickles, R.J. "Improving the yield of 2-(^{18}F)fluoro-2-deoxyglucose using a microwave cavity." *Nucl. Med. Biol.* **1996**, *23*, pp. 605-09.

518. Kad, G.L.; Kaur, J.; Bansal, P.; Singh, J. "Selective iodination of benzylic alcohols with sodium iodide over KSF-clay under microwave irradiation." *J. Chem. Res. (S)* **1996**, pp. 188-89.

519. Hwang, D.R.; Moerlein, S.M.; Lang, L.; Welch, M.J. "Application of microwave technology to the synthesis of short-lived radiopharmaceuticals." *J. Chem. Soc., Chem. Commun.* **1987**, pp. 1799-1801.

520. McCarthy, T.J.; Dence, C.S.; Welch, M.J. "Applications of microwave heating to the synthesis of [^{18}F]-fluoromisonidazole." *Appl. Radiat. Isot.* **1993**, *44*, pp. 1129-32.

521. Banik, B.K.; Becker, F.F. "Synthesis, electrophilic substitution, and structure-activity relationship studies of polycyclic aromatic compounds towards the development of anticancer agents." *Curr. Med. Chem.* **2001**, *8*, pp. 1513-33.

522. Bram, G.; Loupy, A.; Majdoub, M.; Petit, A. "Anthraquinone microwave-induced synthesis in dry media in domestic ovens." *Chem. Ind. (London)* **1991**, pp. 396-97.

523. Marquie, J.; Salmoria, G.; Poux, M.; Laporterie, A.; Dubac, J.; Roques, N. "Acylation and related reactions under microwaves. 5. Development to large laboratory scale with a continuous-flow process." *Ind. Eng. Chem. Res.* **2001**, *40*, pp. 4485-90.

524. Marquie, J.; Laporte, C.; Laporterie, A.; Dubac, J.; Desmurs, J.R.; Roques, N. "Acylation reactions under microwaves. 3. Aroylation of benzene and its slightly activated or deactivated derivatives." *Ind. Eng. Chem. Res.* **2000**, *39*, pp. 1124-31.

525. Laporte, C.; Marquie, J.; Laporterie, A.; Desmurs, J.R.; Dubac, J. "Acylation reactions under microwaves. II. Acylation of aromatic ethers." *C.R. Acad. Sci., Ser. IIc: Chim.* **1999**, *2*, pp. 455-65.

526. Marquie, J.; Laporterie, A.; Dubac, J.; Roques, N.; Desmurs, J.R. "Acylation and related reactions under microwaves. 4. Sulfonylation reactions of aromatics." *J. Org. Chem.* **2001**, *66*, pp. 421-25.

527. Hajipour, A.R.; Mallakpour, S.E.; Imanzadeh, G. "An efficient and novel method for the synthesis of aromatic sulfones under solvent-free conditions." *Indian J. Chem. Sec. B* **2001**, *40*, pp. 237-39.

528. Jones, J.R.; Lockley, W.J.S.; Lu, S.Y.; Thompson, S.P. "Microwave-enhanced aromatic dehalogenation studies: A rapid deuterium-labeling procedure." *Tetrahedron Lett.* **2001**, *42*, pp. 331-32.

529. Srikrishna, A.; Nagaraju, S.; Kondaiah, P. "Application of microwave heating technique for rapid synthesis of γ,β-unsaturated esters." *Tetrahedron* **1995**, *51*, pp. 1809-16.

530. Giraud, L.; Huber, V.; Jenny, T. "2,2-Divinyladamantane: a new substrate for the modification of silicon surfaces." *Tetrahedron* **1998**, *54*, pp. 11899-906.

531. Moghaddam, F.M.; Ghaffarzadeh, M.; Abdi-Oskoui, S.H. "Tandem Fries reaction-conjugate addition under microwave irradiation in dry media; one-pot synthesis to flavanones." *J. Chem. Res. (S)* **1999**, pp. 574-75.

532. Kad, G.L.; Trehan, I.R.; Kaur, J.; Nayyar, S.; Arora, A.; Brar, J.S. "Microwave assisted Fries rearrangement on K10 montmorillonite." *Indian J. Chem.* **1996**, *35B*, pp. 734-36.

533. Sridar, V.; Rao, V.S.S. "Microwave-induced rate enhancement of Fries rearrangement." *Indian J. Chem.* **1994**, *33B*, pp. 184-85.

534. Khadilkar, B.M.; Madyar, V.R. "Fries rearrangement at atmospheric pressure using microwave irradiation." *Synth. Commun.* **1999**, *29*, pp. 1195-1200.

535. Moghaddam, F.M.; Dakamin, M.G. "Thia-Fries rearrangement of aryl sulfonates in dry media under microwave activation." *Tetrahedron Lett.* **2000**, *41*, pp. 3479-81.

536. Heravi, M.M.; Kiakojoori, R.; Mojtahedi, M.M. "Oxidation of alcohols by silica gel-supported ammonium chlorochromate in solventless system." *Indian J. Chem., Sec. B* **2001**, *40*, pp. 329-30.

537. Mojtahedi, M.M.; Saidi, M.R.; Bolourtchian, M.; Shirzi, J.S. "Microwave assisted selective oxidation of benzylic alcohols with calcium hypochlorite under solvent-free conditions." *Monatsh. Chem.* **2001**, *132*, pp. 655-58.

538. Mukhopadhyay, C.; Becker, F.F.; Banik, B.K. "A novel catalytic role of molecular iodine in the oxidation of benzylic alcohols: microwave-assisted reaction." *J. Chem. Res. (S)* **2001**, pp. 28-31.

539. Hajipour, A.R.; Mohammadpoor-Baltork, I. "Solid-phase oxidation of organic compounds with benzyltriphenylphosphonium dichromate." *Phosphorus Sulfur Silicon Rel. Elem.* **2000**, *164*, pp. 145-51.

540. Hajipour, A.R.; Mallakpour, S.E.; Adibi, H. "Oxidation of alcohols with benzyltriphenylphosphonium peroxymonosulfate under non-aqueous conditions." *Phosphorus Sulfur Silicon Rel. Elem.* **2000**, *164*, pp. 71-79.

541. Singh, J.; Sharma, M.; Chhibber, M.; Kaur, J.; Kad, G.L. "Chemoselective oxidation of benzylic alcohols with solid supported CrO_3/TBHP under microwave irradiation." *Synth. Commun.* **2000**, *30*, pp. 3941-45.

542. Tajbakhsh, M.; Ghaemi, M.; Sarabi, S.; Ghassemzadeh, M.; Heravi, M.M. "N-methyl piperidinium chlorochromate adsorbed on alumina: a new and selective reagent for the oxidation of benzylic alcohols to their corresponding carbonyl compounds." *Monatsh. Chem.* **2000**, *131*, pp. 1213-16.

543. Chakraborty, V.; Bordoloi, M. "Microwave assisted oxidation of alcohols by pyridinium chlorochromate." *J. Chem. Res. (S)* **1999**, pp. 118-19.

544. Heravi, M.M.; Ajami, D.; Tabar-Hydar, K.; Ghassemzadeh, M. "Remarkable fast microwave-assisted zeolite HZSM-5 catalyzed oxidation of alcohols with chromium trioxide under solvent-free conditions." *J. Chem. Res. (S)* **1999**, pp. 334-35.

545. Heravi, M.M.; Ajami, D.; Noushabadi, A.M. "Hexamethylenetetramine-bromine on wet alumina: rapid oxidation of alcohols to carbonyl compounds in solventless system using microwaves." *Iranian J. Chem. & Chem. Eng., Intl. Engl.* **1999**, *18*, pp. 88-90.

546. Varma, R.S.; Dahiya, R. "Copper(II) nitrate on clay (clay-cop) - hydrogen peroxide: selective and solvent-free oxidations using microwaves." *Tetrahedron Lett.* **1998**, *39*, pp. 1307-08.

547. Varma, R.S.; Saini, R.K.; Dahiya, R. "Selective oxidations using alumina-supported iodobenzene diacetate under solvent-free conditions." *J. Chem. Res. (S)* **1998**, pp. 324-25.

548. Varma, R.S.; Saini, R.K. "Wet alumina supported chromium(VI) oxide: selective oxidation of alcohols in solventless system." *Tetrahedron Lett.* **1998**, *39*, pp. 1481-82.

549. Varma, R.S.; Saini, R.K.; Dahiya, R. "Active manganese dioxide on silica: oxidation of alcohols under solvent-free conditions using microwaves." *Tetrahedron Lett.* **1997**, *38*, pp. 7823-24.

550. Varma, R.S.; Dahiya, R.; Saini, R.K. "Iodobenzene diacetate on alumina: rapid oxidation of alchohols to carbonyl compounds in solventless system using microwaves." *Tetrahedron Lett.* **1997**, *38*, pp. 7029-32.

551. Varma, R.S.; Dahiya, R. "Microwave-assisted oxidation of alcohols under solvent-free conditions using clayfen." *Tetrahedron Lett.* **1997**, *38*, pp. 2043-44.

552. Yang, D.T.C.; Zhang, C.J.; Haynie, B.C.; Fu, P.P.; Kabalka, G.W. "Microwave assisted oxidation of a-substituted carbonyl compounds to carboxylic acids in aqueous media." *Synth. Commun.* **1997**, *27*, pp. 3235-39.

553. Meng, Q.H.; Feng, J.C.; Bian, N.S.; Liu, B.; Li, C.C. "Benzimadazolium dichromate - a new reagent for selective oxidation under microwave irradiation." *Synth. Commun.* **1998**, *28*, pp. 1097-102.

554. Heravi, M.M.; Kiakojoori, R.; Hydar, K.T. "Ammonium chlorochromate adsorbed on montmorillonite K-10: selective oxidation of alcohols under solvent-free conditions." *J. Chem. Res. (S)* **1998**, pp. 656-57.

555. Raner, K.D.; Strauss, C.R.; Trainor, R.W.; Thorn, J.S. "A new microwave reactor for batchwise organic synthesis." *J. Org. Chem.* **1995**, *60*, pp. 2456-60.

556. Heravi, M.M.; Ajami, D.; Tabar-Heydar, K. "Oxidation of alcohols by silica-supported bis(trimethylsilyl)chromate under microwave irradiation without solvent." *Synth. Commun.* **1999**, *29*, pp. 163-66.

557. Heravi, M.M.; Ajami, D.; Aghapoor, K.; Ghassemzadeh, M. "'Zeofen', a user-friendly oxidizing reagent." *J. Chem. Soc., Chem. Commun.* **1999**, pp. 833-34.

558. Mirza-Aghayan, M.; Heravi, M.M. "Chromium trioxide on H-Y zeolite: rapid oxidation of alcohols to carbonyl compounds in a solventless system using microwaves." *Synth. Commun.* **1999**, *29*, pp. 785-89.

559. Heravi, M.M.; Aghayan, M.M. "Microwave-assisted oxidation of alcohols using wet alumina supported chromium chlorochromate in a solventless system." *Z. Naturforsch., B: Chem. Sci.* **1999**, *54*, pp. 815-17.

560. Palombi, L.; Bonadies, F.; Scettri, A. "Microwave-assisted oxidation of saturated and unsaturated alcohols with *t*-butyl hydroperoxide and zeolites." *Tetrahedron* **1997**, *53*, pp. 15867-876.

561. Hajipour, A.R.; Mallakpour, S.E.; Khoee, S. "An efficient, fast, and selective oxidation of aliphatic and benzylic alcohols to the corresponding carbonyl compounds under microwave irradiation." *Synlett.* **2000**, *5*, pp. 740-42.

562. Khadilkar, B.M.; Gadre, S.A.; Makwana, V.D.; Madyar, V.R. "An efficient and easy preparation of manganese dioxide-silica supported oxidant." *Indian J. Chem.* **1998**, *37A*, pp. 189-90.

563. Bogdal, D.; Lukasiewicz, M. "Microwave-assisted oxidation of alcohols using aqueous hydroperoxide." *Synlett.* **2000**, *1*, pp. 143-45.

564. Balalaie, S.; Golizeh, M.; Hashtroudi, M.S. "Clean oxidation of benzoins on zeolite A using microwave irradiation under solvent-free conditions." *Green Chem.* **2000**, *2*, pp. 277-78.

565. Varma, R.S.; Kumar, D.; Dahiya, R. "Solid state oxidation of benzoins on alumina-supported copper(II) sulfate under microwave irradiation." *J. Chem. Res. (S)* **1998**, pp. 120-21.

566. Varma, R.S.; Dahiya, R.; Kumar, D. "Solvent-free oxidation of benzoins using oxone on wet alumina under microwave irradiation." *Molecules Online* **1998**, *2*, pp. 82-85.

567. Mitra, A.K.; De, A.; Karchaudhuri, N. "Microwave-assisted syntheses of 1,2-diketones." *J. Chem. Res. (S)* **1999**, pp. 246-47.

568. Oussaid, A.; Loupy, A. "Selective oxidation of arenes in dry media under focused microwaves." *J. Chem. Res. (S)* **1997**, pp. 342-43.

569. Varma, R.S.; Saini, R.K.; Meshram, H.M. "Selective oxidation of sulfides to sulfoxides and sulfones by microwave thermolysis on wet silica-supported sodium periodate." *Tetrahedron Lett.* **1997**, *38*, pp. 6525-28.

570. Lie Ken Jie, M.S.F.; Yan-Kit, C. "The use of a microwave oven in the chemical transformation of long chain fatty acid esters." *Lipids* **1988**, *23*, pp. 367-69.

571. Sharifi, A.; Bolourtchian, M.; Mohsenzadeh, F. "Microwave promoted oxidation of α,β-unsaturated ketones in aqueous sodium perborate." *J. Chem. Res. (S)* **1998**, pp. 668-69.

572. Sala, G.D.; Giordano, L.; Lattanzi, A.; Proto, A.; Scettri, A. "Metallocene-catalyzed stereoselective epoxidation of allylic alcohols." *Tetrahedron* **2000**, *56*, pp. 3567-73.

573. Feng, J.C.; Liu, B.; Dai, L.; Yang, X.L.; Tu, S.J. "Microwave assisted solid reaction: reduction of esters to alcohols by potassium borohydride-lithium chloride." *Synth. Commun.* **2001**, *31*, pp. 1875-77.

574. Hajipour, A.R.; Mallakpour, S.E. "Butyltriphenyl-phosphonium tetrahydroborate (BTPPTB) as a selective reducing agent for reduction of organic compounds." *Synth. Commun.* **2001**, *31*, pp. 1177-85.

575. Barbry, D.; Torchy, S. "Accelerated reduction of carbonyl compounds under microwave irradiation." *Tetrahedron Lett.* **1997**, *38*, pp. 2959-60.

576. Varma, R.S.; Saini, R.K. "Microwave-assisted reduction of carbonyl compounds in solid state using sodium borohydride supported on alumina." *Tetrahedron Lett.* **1997**, *38*, pp. 4337-38.

577. Bagnell, L.; Strauss, C.R. "Uncatalyzed hydrogen-transfer reductions of aldehydes and ketones." *J. Chem. Soc., Chem. Commun.* **1999**, pp. 287-88.

578. Erb, W.T.; Jones, J.R.; Lu, S.Y. "Microwave enhanced deuteriations in the solid state using alumina doped sodium borodeuteride." *J. Chem. Res. (S)* **1999**, pp. 728-29.

579. Chen, S.T.; Yu, H.M.; Chen, S.T.; Wang, K.T. "Microwave assisted solid reaction: reduction of ketones using sodium borohydride." *J. Chin. Chem. Soc.* **1999**, *46*, pp. 509-11.

580. Loupy, A.; Monteux, D.; Petit, A.; Merienne, C.; Aizpurua, J.M.; Palomo, C. "Leuckart reductive amination of a 4-acetylazetidinone using microwave technology." *J. Chem. Res. (S)* **1998**, pp. 187, 915-21.

581. Loupy, A.; Monteux, D.; Petit, A.; Aizpurua, J.M.; Dominguez, E.; Palomo, C. "Towards the rehabilitation of the Leuckart reductive amination reaction using microwave technology." *Tetrahedron Lett.* **1996**, *37*, pp. 8177-80.

582. Varma, R.S.; Dahiya, R. "Sodium borohydride on wet clay: solvent-free reductive amination of carbonyl compounds using microwaves." *Tetrahedron* **1998**, *54*, pp. 6293-98.

583. Vass, A.; Dudas, J.; Toth, J.; Varma, R.S. "Solvent-free reduction of aromatic nitro compounds with alumina-supported hydrazine under microwave irradiation." *Tetrahedron Lett.* **2001**, *42*, pp. 5347-49.

584. Bose, A.K.; Manhas, M.S.; Banik, B.K.; Robb, E.W. "Microwave-induced Organic Reaction Enhancement (MORE) chemistry: techniques for rapid, safe, and inexpensive synthesis." *Res. Chem. Intermed.* **1994**, *20*, pp. 1-11.

585. Al-Qahtani, M.H.; Cleator, N.; Danks, T.N.; Garman, R.N.; Jones, J.R.; Stefaniak, S.; Morgan, A.D.; Simmonds, A.J. "Microwave enhanced hydrogenation reactions using solid hydrogen, deuterium, and tritium donors." *J. Chem. Res. (S)* **1998**, pp. 400-01.

586. Dayal, B.; Ertel, N.H.; Rapole, K.R.; Askaongar, A.; Salen, G. "Rapid hydrogenation of unsaturated sterols and bile alcohols using microwaves." *Steroids* **1997**, *62*, pp. 451-54.

587. Banik, B.K.; Barakat, K.J.; Wagle, D.R.; Manhas, M.S.; Bose, A.K. "Microwave-assisted rapid and simplified hydrogenation." *J. Org. Chem.* **1999**, *64*, pp. 5746-53.

588. Gadhwal, S.; Baruah, M.; Sandhu, J.S. "Microwave induced synthesis of hydrazones and Wolff-Kishner reduction of carbonyl compounds." *Synlett.* **1999**, *10*, pp. 1573-74.

589. Parquet, E.; Lin, Q. "Microwave-assisted Wolff-Kishner reduction reaction." *J. Chem. Educ.* **1997**, *74*, p. 1225.

590. Loupy, A.; Song, S.J.; Sohn, S.M.; Lee, Y.M.; Kwon, T.W. "Solvent-free bentonite-catalyzed condensation of malonic acid and aromatic aldehydes under microwave irradiation." *J. Chem. Soc., Perkin Trans. 1* **2001**, pp. 1220-22.

591. Tanaka, K.; Shiraishi, R. "Clean and efficient condensation reactions of aldehydes and amines in a water suspension medium." *Green Chem.* **2000**, *2*, pp. 272-73.

592. Yu, J.H.; Hu, Y.L.; Huang, Q.B.; Ma, R.H.; Yang, S.Y. "A microwave promoted new condensation reaction of aryl ketones with triethyl orthoformate." *Synth. Commun.* **2000**, *30*, pp. 2801-06.

593. Abenhaim, D.; Son, C.P.N.; Loupy, A.; Nguyen, B.H. "Synthesis of jasminaldehyde by solid-liquid phase transfer catalysis without solvent, under microwave irradiation." *Synth. Commun.* **1994**, *24*, pp. 1199-205.

594. Villemin, D.; Martin, B.; Puciova, M.; Toma, S. "Dry synthesis under microwave irradiation: synthesis of ferrocenylenones." *J. Organomet. Chem.* **1994**, *484*, pp. 27-31.

595. Gupta, R.; Gupta, A.K.; Paul, S.; Kachroo, P.L. "Improved microwave-induced synthesis of chalcones and related enones." *Indian J. Chem.* **1995**, *34B*, pp. 61-62.

596. LeGall, E.; Texier-Boullet, F.; Hamelin, J. "Simple access to α,β-unsaturated ketones by acid-catalyzed solvent-free reactions." *Synth. Commun.* **1999**, *29*, pp. 3651-57.

597. Kad, G.L.; Kaur, K.P.; Singh, V.; Singh, J. "Microwave induced rate enhancement in aldol condensation." *Synth. Commun.* **1999**, *29*, pp. 2583-86.

598. Kad, G.L.; Singh, V.; Khurana, A.; Singh, J. "The synthesis of phycopsisenone, a new phenolic secondary metabolite from the sponge, *Phycopsis* sp." *J. Nat. Prod.* **1998**, *61*, pp. 297-98.

599. Babu, G.; Perumal, P.T. "Convenient synthesis of α,α(1)-bis(substituted furfurylidene) cycloalkanones and chalcones under microwave irradiation." *Synth. Commun.* **1997**, *27*, pp. 3677-82.

600. Zheng, M.; Wang, L.; Shao, J.; Zhong, Q. "A facile synthesis of α'-bis(substituted benzylidene)cycloalkanones catalyzed by bis(*p*-ethoxyphenyl)telluroxides (BMPTO) under microwave irradiation." *Synth. Commun.* **1997**, *27*, pp. 351-54.

601. Elder, J.W. "Microwave synthesis of tetraphenylcyclopentadienone and dimethyl tetraphenylphthalate." *J. Chem. Educ.* **1994**, *71*, pp. A142-A144.

602. Balalaie, S.; Nemati, N. "One-pot preparation of coumarins by Knoevenagel condensation in solvent-free condition under microwave irradiation." *Heterocycl. Commun.* **2001**, *7*, pp. 67-72.

603. Reddy, G.V.; Maitraie, D.; Narsaiah, B.; Rambabu, Y.; Rao, P.S. "Microwave assisted Knoevenagel condensation: a facile method for the synthesis of chalcones." *Synth. Commun.* **2001**, *31*, pp. 2881-84.

604. Dave, C.G.; Augustine, C. "Microwave assisted Knoevenagel condensation using NaCl and NH_4OAc-AcOH system as catalysts under solvent-free conditions." *Indian J. Chem., Sect. B* **2000**, *39*, pp. 403-05.

605. Bogdal, D. "Coumarins - fast synthesis by the Knoevenagel condensation under microwave irradiation." *J. Chem. Res. (S)* **1998**, pp. 468-69.

606. Kim, S.Y.; Kwon, P.S.; Kwon, T.W.; Chung, S.K.; Chang, Y.T. "Microwave enhanced Knoevenagel condensation of ethyl cyanoacetate with aldehydes." *Synth. Commun.* **1997**, *27*, pp. 533-41.

607. de la Cruz, P.; Diez-Barra, E.; Loupy, A.; Langa, F. "Silica gel catalyzed Knoevenagel condensation in dry media under microwave irradiation." *Tetrahedron Lett.* **1996**, *37*, pp. 1113-16.

608. Peng, Y.Q.; Song, G.H.; Qian, X.H. "Urotropine: an efficient catalyst precursor for the microwave-assisted Knoevenagel reaction." *J. Chem Res. (S)* **2001**, pp. 188-89.

609. Kuster, G.J.; Scheeren, H.W. "The preparation of resin-bound nitroalkenes and some applications in high pressure promoted cycloadditions." *Tetrahedron Lett.* **2000**, *41*, pp. 515-19.

610. Mitra, A.K.; De, A.; Karchaudhuri, N. "Solvent-free microwave enhanced Knoevenagel condensation of ethyl cyanoacetate with aldehydes." *Synth. Commun.* **1999**, *29*, pp. 2731-39.

611. Abdallah-El Ayoubi, S.; Texier-Boullet, F.; Hamelin, J. "Minute synthesis of electrophilic alkenes under microwave irradiation." *Synthesis* **1994**, *3*, pp. 258-60.

612. Sabitha, G.; Reddy, B.V.S.; Babu, S.R.; Yadav, J.S. "LiCl catalyzed Knoevenagel condensation: comparative study of conventional method vs. microwave irradiation." *Chem. Lett.* **1998**, pp. 773-74.

613. Balalaie, S.; Nemati, N. "Ammonium acetate-basic alumina catalyzed Knoevenagel condensation under microwave irradiation under solvent-free condition." *Synth. Commun.* **2000**, *30*, pp. 869-75.

614. Kwon, P.S.; Kim, Y.M.; Kang, C.J.; Kwon, T.W.; Chung, S.K.; Chang, Y.T. "Microwave enhanced Knoevenagel condensation of malonic acid on basic alumina." *Synth. Commun.* **1997**, *27*, pp. 4091-100.

615. Abdallah-El Ayoubi, S.; Texier-Boullet, F. "Clay-mediated synthesis of gem-bis(alkoxycarbonyl)alkenes under microwave irradiation." *J. Chem. Res. (S)* **1995**, pp. 208-09.

616. Kim, J.K.; Kwon, P.S.; Kwon, T.W.; Chung, S.K.; Lee, J.W. "Application of microwave irradiation techniques for the Knoevenagel condensation." *Synth. Commun.* **1996**, *26*, pp. 535-42.

617. Gasparova, R.; Lacova, M. "Study of microwave irradiation effect on condensation of 6-R-3-formylchromones with active methylene compounds." *Collect. Czech. Chem. Commun.* **1995**, *60*, pp. 1178-85.

618. Kumar, H.M.S.; Reddy, B.V.S.; Anjaneyulu, S.; Yadav, J.S. "Non solvent reaction: ammonium acetate catalyzed highly convenient preparation of *trans*-cinnamic acids." *Synth. Commun.* **1998**, *28*, pp. 3811-15.

619. Mitra, A.K.; De, A.; Karchaudhuri, N. "Application of microwave irradiation techniques to the syntheses of cinnamic acids by Doebner condensation." *Synth. Commun.* **1999**, *29*, pp. 573-81.

620. Kumar, H.M.S.; Reddy, B.V.S.; Reddy, E.J.; Yadav, J.S. "SiO_2 catalyzed expedient synthesis of (E)-3-alkenoic acids in dry media." *Tetrahedron Lett.* **1999**, *40*, pp. 2401-04.

621. Villemin, D. Martin, B. "Clay catalysis: an easy synthesis of 5-nitrofuraldehyde and 5-nitrofurfurylidene derivatives under microwave irradiation." *J. Chem. Res. (S)* **1994**, pp. 146-47.

622. Villemin, D. Martin, B. "Potassium fluoride on alumina: dry synthesis of 3-arylidene-1,3-dihydro-indol-2-one under microwave irradiation." *Synth. Commun.* **1998**, *28*, pp. 3201-08.

623. Villemin, D.; Martin, B.; Bar, N. "Application of microwave in organic synthesis. Dry synthesis of 2-arylmethylene-3(2)-naphthofuranones." *Molecules* **1998**, *3*, pp. 88-93.

624. Villemin, D.; Martin, B.; Khalid, M. "Dry reaction on KF-alumina: synthesis of 4-arylidene-1,3-(2H,4H)-isoquinolinediones." *Synth. Commun.* **1998**, *28*, pp. 3195-200.

625. Frere, S.; Thiery, V.; Besson, T. "Microwave acceleration of the Pechmann reaction on graphite/montmorillonite K-10: application to the preparation of 4-substituted-7-amino-coumarins." *Tetrahedron Lett.* **2001**, *42*, pp. 2791-94 and *Fifth International Electronic Conference on Synthetic Organic Chemistry (ECSOC-5)* **2001**, E0015 (www.mdpi.net).

626. Varma, R.S.; Dahiya, R.; Kumar, S. "Microwave-assisted Henry reactions: solventless synthesis of conjugated nitroalkenes." *Tetrahedron Lett.* **1997**, *38*, pp. 5131-34.

627. Kumar, H.M.S.; Reddy, B.V.S.; Yadav, J.S. "SiO_2 catalyzed Henry reaction: microwave assisted preparation of 2-nitroalkanols in dry media." *Chem. Lett.* **1998**, pp. 637-38.

628. Kabalka, G.W.; Wang, L.; Pagni, R.M. "A microwave-enhanced, solventless Mannich condensation on CuI-doped alumina." *Synlett.* **2001**, pp. 676-78.

629. Kabalka, G.W.; Wang, L.; Pagni, R.M. "A novel route to 2-(dialkylaminomethyl) benzo[*b*]furans via a microwave-enhanced, solventless Mannich condensation-cyclization on cuprous iodide doped alumina." *Tetrahedron Lett.* **2001**, *42*, pp. 6049-51.

630. Gadhwal, S.; Baruah, M.; Prajapati, D.; Sandhu, J.S. "Microwave-assisted regioselective synthesis of β-amino ketones via the Mannich reaction." *Synlett.* **2000**, *3*, pp. 341-42.

631. Hoel, A.M.L.; Nielsen, J. "Microwave-assisted solid-phase Ugi four-component condensations." *Tetrahedron Lett.* **1999**, *40*, pp. 3941-44.

632. Li, K.L.; Xia, L.X.; Li, J.; Pang, J.; Cao, G.Y.; Xi, Z.W. "Salt-assisted acid hydrolysis of starch to D-glucose under microwave irradiation." *Carbohydrate Research* **2001**, *331*, pp. 9-12.

633. Moghaddam, F.M.; Ghaffarzadeh, M. "Microwave-assisted rapid hydrolysis and preparation of thioamides by Willgerodt-Kindler reaction." *Synth. Commun.* **2001**, *31*, pp. 317-21.

634. Dayal, B.; Ertel, N.H. "Rapid hydrolysis of bile acid conjugates using microwaves: retention of absolute stereochemistry in the hydrolysis of (25R) 3-α, 7-α, 12-α-trihydroxy-5-β-cholestan-26-oyltaurine." *Lipids* **1998**, *33*, pp. 333-38.

635. Stenberg, M.; Marko-Varga, G.; Oste, R. "Racemization of amino acids during classical and microwave oven hydrolysis - application to aspartame and a Maillard reaction system." *Food Chem.* **2001**, *74*, pp. 217-24.

636. Ranu, B.C.; Dutta, P.; Sarkar, A. "An efficient and general method for ester hydrolysis on the surface of silica gel catalyzed by indium triiodide under microwave irradiation." *Synth. Commun.* **2000**, *30*, pp. 4167-71.

637. Kabza, K.G.; Gestwicki, J.E.; McGrath, J.L.; Petrassi, H.M. "Effect of microwave radiation on copper(II) 2,2'-bipyridyl-mediated hydrolysis of bis(*p*-nitrophenyl) phosphodiester and enzymatic hydrolysis of carbohydrates." *J. Org. Chem.* **1996**, *61*, pp. 9599-602.

638. Gedye, R.N.; Smith, F.E.; Westaway, K.C. "The rapid synthesis of organic compounds in microwave ovens." *Can. J. Chem.* **1988**, *66*, pp. 17-26.

639. Das, B.; Venkataiah, B.; Kashinatham, A. "Chemical and biochemical modifications of parthenin." *Tetrahedron* **1999**, *55*, pp. 6585-94.

640. Subbaraju, G.V.; Manhas, M.S.; Bose, A.K. "Studies on terpenoids. An improved synthesis of (-)-dihydroactinidiolide." *Tetrahedron Lett.* **1991**, *32*, pp. 4871-74.

641. Das, B.; Venkataiah, B. "Synthetic studies on natural products. 7. Conversion of parthenin to anhydroparthenin using microwave irradiation." *Synth. Commun.* **1999**, *29*, pp. 863-66.

642. Jayaraman, M.; Batista, M.T.; Manhas, M.S.; Bose, A.K. "Studies of lactams. 106. Organic reactions in water: indium mediated synthesis of α-alkylidene-β-lactams." *Heterocycles* **1998**, *49*, pp. 97-100.

643. Meshram, H.M.; Sekhar, K.C.; Ganesh, Y.S.S.; Yadav, J.S. "Microwave thermolysis. XI. Clay catalyzed facile cyclodehydration under microwave: synthesis of 3-substituted benzofurans." *Synlett.* **2000**, pp. 1273-74.

644. Varma, R.S.; Saini, R.K.; Kumar, D. "An expeditious synthesis of flavones on Montmorillonite K-10 clay with microwaves." *J. Chem. Res. (S)* **1998**, pp. 348-49.

645. Kumar, H.M.S.; Mohanty, P.K.; Kumar, M.S.; Yadav, J.S. "Microwave promoted rapid dehydration of aldoximes to nitriles on a solid support." *Synth. Commun.* **1997**, *27*, pp. 1327-33.

646. Sabitha, G.; Syamala, M. "Microwave induced conversion of aldozimes to nitriles by DBU." *Synth. Commun.* **1998**, *28*, pp. 4577-80.

647. Das, B.; Madhusudhan, P.; Venkataiah, B. "Studies on novel synthetic methodologies. Part 4. An efficient microwave-assisted one-pot conversion of aldehydes into nitriles using silica gel-supported $NaHSO_4$ catalyst." *Synlett.* **1999**, *10*, pp. 1569-70.

648. Bose, D.S.; Narsaiah, A.V. "Efficient one-pot synthesis of nitriles from aldehydes in solid state using peroxymonosulfate on alumina." *Tetrahedron Lett.* **1998**, *39*, pp. 6533-34.

649. Chakraborti, A.K.; Kaur, G. "One-pot synthesis of nitriles from aldehydes under microwave irradiation: influence of the medium and mode of microwave irradiation on product formation." *Tetrahedron* **1999**, *55*, pp. 13265-268.

650. Feng, J.C.; Liu, B.; Bian, N.S. "One-step conversion of aldehydes into nitriles in dry media under microwave irradiation." *Synth. Commun.* **1998**, *28*, pp. 3765-68.

651. Delgado, F.; Cano, A.C.; Garcia, O.; Alvarado, J.; Velasco, L.; Alvarez, C.; Rudler, H. "A direct synthesis of aromatic nitriles from aldehydes using a Mexican bentonite and microwave or infrared irradiation, in absence of solvent." *Synth. Commun.* **1992**, *22*, pp. 2125-28.

652. Feng, J.C.; Liu, B.; Liu, Y.; Li, C.C. "An efficient one-pot synthesis of nitriles from carboxylic acids without solvent under microwave irradiation." *Synth. Commun.* **1996**, *26*, pp. 4545-48.

653. Cros, E.; Planas, M.; Bardaji, E. "Synthesis of N-α-tetrachlorophthaloyl (TCP)-protected amino acids under microwave irradiation (MWI)." *Synthesis* **2001**, pp. 1313-20.

654. Deka, N.; Sarma, J.C. "Microwave-mediated selective monotetrahydropyranylation of symmetrical diols catalyzed by iodine." *J. Org. Chem.* **2001**, *66*, pp. 1947-48.

655. Khalafi-Nezhad, A.; Alamdari, R.F.; Zekri, N. "Efficient and selective protection of alcohols and phenols with triisopropylsilyl chloride/imidazole using microwave irradiation." *Tetrahedron* **2000**, *56*, pp. 7503-06.

656. Heravi, M.M.; Ajami, D.; Ghassemzadeh, M. "Solvent free tetrahydropyranylation of alcohols and phenols over sulfuric acid adsorbed on silica gel." *Synth. Commun.* **1999**, *29*, pp. 1013-16.

657. Csiba, M.; Cleophax, J.; Loupy, A.; Malthete, J.; Gero, S.D. "Liquid crystalline 5,6-O-acetals of L-galactono-1,4-lactone prepared by microwave irradiation on montmorillonite." *Tetrahedron Lett.* **1993**, *34*, pp. 1787-90.

658. Perio, B.; Hamelin, J. "Oxathiolane and dithiolane exchange reaction for carbonyl group protection: a new, fast and efficient procedure without solvent under microwave irradiation." *Green Chem.* **2000**, *2*, pp. 252-55.

659. Kad, G.L.; Singh, V.; Kaur, K.P.; Singh, J. "Microwave assisted preparation of 1,3-dithiolanes under solvent free conditions." *Indian J. Chem.* **1998**, *37B*, pp. 172-73.

660. Pourjavadi, A.; Mirjalili, B.F. "Microwave-assisted rapid ketalization/acetalization of aromatic aldehydes and ketones in aqueous media." *J. Chem. Res. (S)* **1999**, pp. 562-63.

661. Beregszaszi, T.; Molnar, A. "Microwave-assisted acetalization of carbonyl compounds catalyzed by reusable Envirocat supported reagents." *Synth. Commun.* **1997**, *27*, pp. 3705-09.

662. Moghaddam, F.M.; Sharifi, A. "Microwave promoted acetalization of aldehydes and ketones." *Synth. Commun.* **1995**, *25*, pp. 2457-61.

663. Kalita, D.J.; Borah, R.; Sarma, J.C. "A new selective catalytic acetalization method promoted by microwave irradiation." *Tetrahedron Lett.* **1998**, *39*, pp. 4573-74.

664. Wang, C.D.; Shi, X.Z.; Xie, R.J. "Synthesis of diacetals from 2,2-bis(hydroxymethyl)-1,3-propanediol under microwave irraditation." *Synth. Commun.* **1997**, *27*, pp. 2517-20.

665. Perio, B.; Dozias, M.J.; Jacquault, P.; Hamelin, J. "Solvent free protection of carbonyl group under microwave irradiation." *Tetrahedron Lett.* **1997**, *38*, pp. 7867-70.

666. Nahar, P. "Microwaves - a powerful tool for the base protection of cytidine." *Tetrahedron Lett.* **1997**, *38*, pp. 7253-54.

667. Yadav, J.S.; Reddy, B.V.S.; Srinivas, R.; Ramalingam, T. "Silica gel-supported metallic sulfates catalyzed chemoselective acetalization of aldehydes under microwave irradiation." *Synlett.* **2000**, *5*, pp. 701-03.

668. Kulkarni, P.P.; Kadam, A.J.; Mane, R.B.; Desai, U.V.; Wadgaonkar, P.P. "Demethylation of methyl aryl ethers using pyridine hydrochloride in solvent-free conditions under microwave irradiation." *J. Chem. Res. (S)* **1999**, pp. 394-95.

669. Varma, R.S.; Varma, M.; Chatterjee, A.K. "Microwave-assisted deacetylation on alumina: a simple deprotection method." *J. Chem. Soc., Perkin Trans. 1* **1993**, pp. 999-1000.

670. Perez, E.R.; Marrero, A.L.; Perez, R.; Autie, M.A. "An efficient microwave-assisted method to obtain 5-nitrofurfural without solvents on mineral solid supports." *Tetrahedron Lett.* **1995**, *36*, pp. 1779-82.

671. Heravi, M.M.; Ajami, D.; Ghassemzadeh, M. "Wet alumina supported chromium(VI) oxide: a mild, efficient, and inexpensive reagent for oxidative deprotection of trimethylsilyl and tetrahydropyranyl ethers in solventless systems." *Synth. Commun.* **1999**, *29*, pp. 781-84.

672. Heravi, M.M.; Ajami, D. "Clay-supported bis(trimethylsilyl)-chromate. Oxidative deprotection of tetrahydropyranyl ethers under solvent-free conditions using microwaves." *Monatsh. Chem.* **1999**, *130*, pp. 709-12.

673. Oussaid, A.; Thach, L.N.; Loupy, A. "Selective dealkylation of alkyl aryl ethers in heterogeneous basic media under microwave irradiation." *Tetrahedron Lett.* **1997**, *38*, pp. 2451-54.

674. Yadav, J.S.; Meshram, H.M.; Reddy, G.S.; Sumithra, G. "Microwave thermolysis. Part 4: Selective deprotection of MPM ethers using clay supported ammonium nitrate 'clayan' in dry media." *Tetrahedron Lett.* **1998**, *39*, pp. 3043-46.

675. Heravi, M.M.; Hekmatshoar, R.; Beheshtiha, Y.S.; Ghassemzadeh, M. "Ammonium chlorochromate adsorbed on montmorillonite K-10: oxidative deprotection of tetrahydropyranyl ethers using microwaves in a solventless system." *Monatsh. Chem.* **2001**, *132*, pp. 651-54.

676. Heravi, M.M.; Ajami, D.; Mojtahedi, M.M.; Ghassemzadeh, M. "A convenient oxidative deprotection of tetrahydropyranyl ethers with iron(III) nitrate and clay under microwave irradiation in solvent free conditions." *Tetrahedron Lett.* **1999**, *40*, pp. 561-62.

677. Heravi, M.M.; Beheshtiha, Y.S.; Oskooi, S.H.A.; Shoar, R.H.; Khalilpoor, M. "Direct oxidative deprotection using montmorillonite supported ammonium chlorochromate under conventional heating and microwave irradiation in solventless system." *Phosphorus Sulfur Silicon Rel. Elem.* **2000**, *161*, pp. 251-55.

678. Heravi, M.M.; Ajami, D.; Ghassemzadeh, M.; Tabar-Hydar, K. "Zeofen, an efficient reagent for oxidative deprotection of trimethylsilyl ethers under microwave irradiation in solventless system." *Synth. Commun.* **2001**, *31*, pp. 2097-100.

679. Varma, R.S.; Lamture, J.B.; Varma, M. "Alumina-mediated cleavage of *t*-butyldimethyl-silyl ethers." *Tetrahedron Lett.* **1993**, *34*, pp. 3029-32.

680. Hajipour, A.R.; Mallakpour, S.E.; Baltork, I.M.; Adibi, H. "Oxidative deprotection of trimethylsilyl ethers, tetrahydropyranyl ethers, and ethylene acetals with benzyltriphenylphosphonium peroxymonosulfate under microwave irradiation." *Synth. Commun.* **2001**, *31*, pp. 1625-31.

681. Mojtahedi, M.M.; Saidi, M.R.; Heravi, M.M.; Bolourtchian, M. "Microwave assisted deprotection of trimethylsilyl ethers under solvent-free conditions catalyzed by clay or a palladium complex." *Monatsh. Chem.* **1999**, *130*, pp. 1175-78.

682. Mojtahedi, M.M.; Saidi, M.R.; Bolourtchian, M.; Heravi, M.M. "Solid state oxidative deprotection of trimethylsilyl ethers with iron(III)nitrate and montmorillonite under microwave irradiation." *Synth. Commun.* **1999**, *29*, pp. 3283-87.

683. Chavan, S.P.; Soni, P.; Kamat, S.K. "Chemoselective deprotection of 1,3-oxathiolanes using Amberlyst 15 and glyoxylic acid under solvent free conditions." *Synlett.* **2001**, pp. 1251-52.

684. Varma, R.S.; Saini, R.K. "Solid state dethioacetalization using clayfen." *Tetrahedron Lett.* **1997**, *38*, pp. 2623-24.

685. Meshram, H.M.; Reddy, G.S.; Sumitra, G.; Yadav, J.S. "Microwave thermolysis. VI. A rapid and general method for dethioacetalization using "Clayan" in dry media." *Synth. Commun.* **1999**, *29*, pp. 1113-19.

686. Meshram, H.M.; Sumitra, G.; Reddy, G.S.; Ganesh, Y.S.S.; Yadav, J.S. "Microwave thermolysis. V. A rapid and selective method for the cleavage of THP ethers, acetals, and acetonides using clay supported ammonium nitrate "Clayan" in dry media." *Synth. Commun.* **1999**, *29*, pp. 2807-15.

687. Bose, D.S.; Jayalakshmi, B.; Narsaiah, A.V. "Efficient method for selective cleavage of acetals and ketals using peroxymonosulfate on alumina under solvent-free conditions." *Synthesis* **2000**, *1*, pp. 67-68.

688. Ballini, R.; Bordoni, M.; Bosica, G.; Maggi, R.; Sartori, G. "Solvent free synthesis and deprotection of 1,1-diacetates over a commercially available zeolite Y as a reusable catalyst." *Tetrahedron Lett.* **1998**, *39*, p. 7587.

689. Varma, R.S.; Chatterjee, A.K.; Varma, M. "Alumina-mediated deacetylation of benzaldehyde diacetates. A simple deprotection method." *Tetrahedron Lett.* **1993**, *34*, pp. 3207-10.

690. Boruah, A.; Baruah, B.; Prajapati, D.; Sandhu, J.S. "Bi(III)-catalyzed regeneration of carbonyl compounds from hydrazones under microwave irradiation." *Synlett.* **1997**, *11*, pp. 1251-52.

691. Mitra, A.K.; De, A.; Karchaudhuri, N. "Regeneration of ketones from semicarbazones in the solid state on wet silica supported sodium bismuthate under microwave irradiation." *J. Chem. Res. (S)* **1999**, pp. 320-21.

692. Varma, R.S.; Meshram, H.M. "Solid state cleavage of semicarbazones and phenylhydrazones with ammonium persulfate-clay using microwave or ultrasonic irradiation." *Tetrahedron Lett.* **1997**, *38*, pp. 7973-76.

693. Baruah, B.; Prajapati, D.; Sandhu, J.S. "Regeneration of carbonyl compounds from semicarbazones under microwave irradiations." *Synth. Commun.* **1998**, *28*, pp. 4157-63.

694. Meshram, H.M.; Srinivas, D.; Reddy, G.S.; Yadav, J.S. "Clay supported ammonium nitrate, "Clayan": a rapid and convenient regeneration of carbonyls in dry media." *Synth. Commun.* **1998**, *28*, pp. 4401-08.

695. Mitra, A.K.; De, A.; Karchaudhuri, N. "Regeneration of ketones from oximes in the solid state on wet silica supported sodium bismuthate under microwave irradiation." *Synlett.* **1998**, *12*, pp. 1345-46.

696. Bendale, P.M.; Khadilkar, B.M. "Microwave promoted regeneration of carbonyl compounds from oximes using silica supported chromium trioxide." *Tetrahedron Lett.* **1998**, *39*, pp. 5867-68.

697. Boruah, A.; Baruah, B.; Prajapati, D.; Sandhu, J.S. "Regeneration of carbonyl compounds from oximes under microwave irradiation." *Tetrahedron Lett.* **1997**, *38*, pp. 4267-68.

698. Varma, R.S.; Meshram, H.M. "Solid state deoximation with ammonium persulfate-silica gel: regeneration of carbonyl compounds using microwaves." *Tetrahedron Lett.* **1997**, *38*, pp. 5427-28.

699. Varma, R.S.; Dahiya, R.; Saini, R.K. "Solid state regeneration of ketones from oximes on wet silica supported sodium periodate using microwaves." *Tetrahedron Lett.* **1997**, *38*, pp. 8819-20.

700. Chakraborty, V.; Bordoloi, M. "Deoximation by pyridinium chlorochromate under microwave irradiation." *J. Chem. Res. (S)* **1999**, pp. 120-21.

701. Gajare, A.S.; Shaikh, N.S.; Bonde, B.K.; Deshpande, V.H. "Microwave accelerated selective and facile deprotection of allyl esters catalyzed by montmorillonite K-10." *J. Chem. Soc., Perkin Trans. 1* **2000**, pp. 639-40.

702. Varma, R.S.; Chatterjee, A.K.; Varma, M. "Alumina-mediated microwave thermolysis: a new approach to deprotection of benzyl esters." *Tetrahedron Lett.* **1993**, *34*, pp. 4603-06.

703. Loupy, A.; Pigeon, P.; Ramdani, M.; Jacquault, P. "Solid-liquid phase-transfer catalysis without solvent coupled with microwave irradiation: a quick and efficient method for saponification of esters." *Synth. Commun.* **1994**, *24*, pp. 159-65.

704. Fabis, F.; Jolivet-Fouchet, S.; Robba, M.; Landelle, H.; Rault, S. "Thiasatoic anhydrides: efficient synthesis under microwave heating conditions and study of their reactivity." *Tetrahedron Lett.* **1998**, *39*, pp. 10789-800.

705. Vass, A.; Tapolcsanyi, P.; Wolfling, J.; Schneider, G. "Microwave-induced selective deacetylation and stereospecific acyl migration of steroid acetates on alumina." *J. Chem. Soc., Perkin Trans. 1* **1998**, pp. 2873-75.

706. An, J.; Bagnell, L.; Cablewski, T.; Strauss, C.R.; Trainor, R.W. "Applications of high-temperature aqueous media for synthetic organic reactions." *J. Org. Chem.* **1997**, *62*, pp. 2505-11.

707. Das, B.; Madhusudhan, P.; Venkataiah, B. "Clay catalysed convenient isomerization of natural furofuran lignans under microwave irradiation." *Synth. Commun.* **2000**, *30*, pp. 4001-06.

708. Loupy, A.; Thach, L.N. "Base-catalyzed isomerization of eugenol: solvent-free conditions and microwave activation." *Synth. Commun.* **1993**, *23*, pp. 2571-77.

709. Loupy, A.; Pigeon, P.; Ramdani, M.; Jacquault, P. "A new solvent-free procedure using microwave technology as an alternative to the Krapcho reaction." *J. Chem. Res. (S)* **1993**, pp. 36-37.

710. Kiddle, J.J. "Microwave irradiation in organophosphorus chemistry. Part 2: Synthesis of phosphonium salts." *Tetrahedron Lett.* **2000**, *41*, pp. 1339-41.

711. Frattini, S.; Quai, M.; Cereda, E. "Kinetic study of microwave-assisted Wittig reaction of stabilized ylides with aromatic aldehydes." *Tetrahedron Lett.* **2001**, *42*, pp. 6827-29.

712. Westman, J. "An efficient combination of microwave dielectric heating and the use of solid-supported triphenylphosphine for Wittig reactions." *Org. Lett.* **2001**, *3*, pp. 3745-48.

713. Chen, M.; Yuan, G.; Yang, S. "A new and facile method for synthesis of aza-Wittig reagents under microwave irradiation." *Synth. Commun.* **2000**, *30*, pp. 1287-94.

714. Murphy, P.J.; Lee, S.E. "Recent synthetic applications of the non-classical Wittig reaction." *J. Chem. Soc., Perkin Trans. 1* 1999, pp. 3049-66.

715. Sabitha, G.; Reddy, M.M.; Srinivas, D.; Yadav, J.S. "Microwave irradiation: Wittig olefination of lactones and amides." *Tetrahedron Lett.* **1999**, *40*, pp. 165-66.

716. Fu, C.; Xu, C.; Huang, Z.Z.; Huang, X. "α,β-Unsaturated sulfones by the Wittig reaction of stable ylide with aldehydes under microwave irradiation." *Org. Prep. Proc. Intl.* **1997**, *29*, pp. 587-89.

717. Spinella, A; Fortunati, T.; Soriente, A. "Microwave accelerated Wittig reactions of stablized phosphorus ylides with ketones under solvent-free conditions." *Synlett.* **1997**, pp. 93-94.

718. Lakhrissi, Y.; Taillefumier, C.; Lakhrissi, M.; Chapleur, Y. "Efficient conditions for the synthesis of C-glycosylidene derivatives: a direct and stereoselective route to C-glycosyl compounds." *Tetrahedron Asymm.* **2000**, *11*, pp. 417-21.

719. Xu, C.; Chen, G.; Fu, C.; Huang, X. "The Wittig reaction of stable ylide and aldehyde under microwave irradiation: synthesis of ethyl cinnamates." *Synth. Commun.* **1995**, *25*, pp. 2229-33.

720. Hachemi, M.; Puciova-Sebova, M.; Toma, S.; Villemin, D. "Dry reaction of diethyl cyanomethylphosphonate tetraethylmethylenediphosphonate with benzaldehyde on solid bases." *Phosphorus Sulfur Silicon Relat. Elem.* **1996**, *113*, pp. 131-36.

721. Latouche, R.; Texier-Boullet, F.; Hamelin, J. "Alkali metal fluoride-mediated silyl-Reformatskii reaction in solid-liquid media. Activation by microwaves." *Tetrahedron Lett.* **1991**, *32*, pp. 1179-82.

722. European Patent No. 0155893, "Appareil de reaction chimique par voie humide de produits divers." **1985**.

723. Gasgnier, M.; Loupy, A.; Petit, A.; Jullien, H. "New developments in the field of energy transfer by means of mono-mode microwaves for various oxides and hydroxides." *J. Alloys Compd.* **1994**, *204*, pp. 165-72.

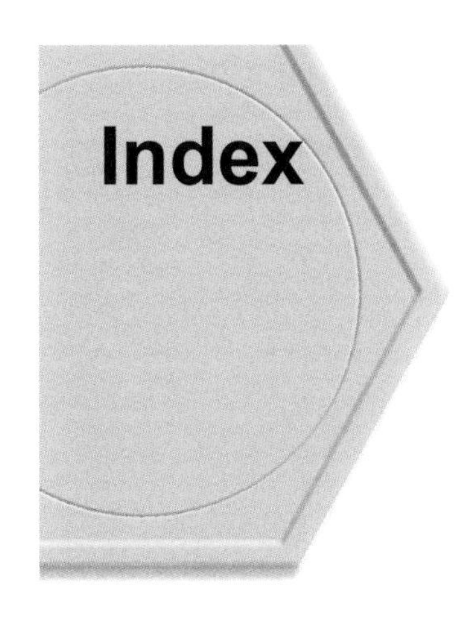

Index

About the Author

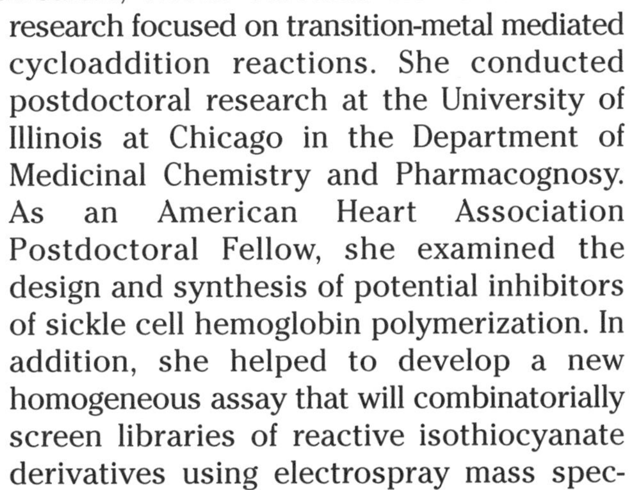

Brittany L. Hayes received her Ph.D. degree in Organic Chemistry from Wake Forest University in Winston-Salem, North Carolina. Her dissertation research focused on transition-metal mediated cycloaddition reactions. She conducted postdoctoral research at the University of Illinois at Chicago in the Department of Medicinal Chemistry and Pharmacognosy. As an American Heart Association Postdoctoral Fellow, she examined the design and synthesis of potential inhibitors of sickle cell hemoglobin polymerization. In addition, she helped to develop a new homogeneous assay that will combinatorially screen libraries of reactive isothiocyanate derivatives using electrospray mass spectrometry (ESMS). Dr. Hayes is a Senior Scientist in the Life Science Division of CEM Corporation.